Analytical Liquid Chromatography - New Perspectives

Edited by Serban C. Moldoveanu and Victor David

Published in London, United Kingdom

Analytical Liquid Chromatography - New Perspectives
http://dx.doi.org/10.5772/intechopen.97995
Edited by Serban C. Moldoveanu and Victor David

Contributors
Ngoc-Van Thi Nguyen, Serban C. Moldoveanu, Victor David, Robert Owen Bussey III, Ayenew Ashenef, Esayas Tesfaye, Tadele Eticha, Ariaya Hymete

First published in London, United Kingdom, 2022 by IntechOpen
IntechOpen is the global imprint of INTECHOPEN LIMITED, registered in England and Wales, registration number: 11086078, 5 Princes Gate Court, London, SW7 2QJ, United Kingdom

British Library Cataloguing-in-Publication Data
A catalogue record for this book is available from the British Library

Additional hard and PDF copies can be obtained from orders@intechopen.com

Analytical Liquid Chromatography - New Perspectives
Edited by Serban C. Moldoveanu and Victor David
p. cm.
Print ISBN 978-1-80355-513-3
Online ISBN 978-1-80355-514-0
eBook (PDF) ISBN 978-1-80355-515-7

Meet the editors

Dr. Serban C. Moldoveanu is a senior principal scientist at R. J. Reynolds Tobacco Company. His research activity is focused on various aspects of chromatography including GC/MS, HPLC, LC/MS/MS, and pyrolysis, with applications mainly to natural products. He is the author of more than 150 original papers, 12 books, and several book chapters. He is a member of the editorial board of *Analytical Methods in Chemistry.*

Professor Victor David was the head of the Analytical Chemistry Department, University of Bucharest, Romania, between 2007 and 2019, and is now an Emeritus Professor. His main field of research is separation science. He is the author of 5 books, 11 chapters in books and encyclopedias, and 150 scientific articles in ISI journals. He was an associate editor of the Journal of Liquid Chromatography & Related Technologies and is a member of the editorial board of *Biomedical Chromatography, Molecules, Journal of Chemistry, Journal of Essential Oil-Bearing Plants, Revue Roumaine de Chimie*, and *Current Chromatography.*

Contents

Preface

This book discusses several aspects of and recent advances in analytical liquid chromatography (LC), referring mainly to high-performance liquid chromatography (HPLC) and ultra-performance liquid chromatography (UPLC). It is not intended to be comprehensive and thus only some selected but important subjects are included. Chapter 1 addresses one such subject, which is the technology of the chromatographic columns used in HPLC and UPLC. The chromatographic column is at the core of HPLC/UPLC separation and significant effort is being made to make improvements to the column. This effort involves improving the type of columns that are widely used in everyday work for practical analyses and developing exploratory new materials for the stationary phase of the columns and miniaturization.

One of the newest developments in HPLC/UPLC is the use of mixed-mode stationary phases, which is discussed in Chapter 2. Mixed-mode phases have at least two different functionalities that can be involved in different types of interaction with analytes. Examples of such different interactions include reversed-phase/hydrophilic interaction, reversed-phase/ion-exchange, hydrophilic interaction/ion-exchange, and even trimodal-type interactions. Different procedures are used to obtain mixed-mode stationary phases, such as mixing particles with different types of active phases, phases containing different ligands, or phases having two (or more) embedded functionalities in the active stationary phase. The chapter discusses the subject of mixed-mode types of stationary phases and presents some of the newest applications of mixed-mode phases in the analysis of pharmaceuticals and biological samples.

Another development in analytical LC is related to multidimensional separations. Although the interest in multidimensional separation in LC is not new, the increasing need for the analysis of complex samples has made the technique more relevant. In practice, the most common type of multidimensional LC is bidimensional, which is examined in Chapter 3. The chapter presents different techniques used in bidimensional LC such as on-line, stop-and-go, and off-line. Also, the chapter describes various applications of multidimensional LC such as in proteomics, lipidomics, environmental analysis, polymer and oligomer separation, metabolomics, and chiral separations.

Chapter 4 is dedicated to the use of ionic liquids in HPLC/UPLC. The chapter discusses stationary phases with attached structures of ionic liquids, the use of ionic liquids as additives in the mobile phase of HPLC separations, and the retention behavior of ionic liquids studied as analytes. It also presents the advantages and shortcomings of using ionic liquids, and the reasons ionic liquids are more and more frequently used in HPLC.

Finally, Chapter 5 discusses the ion mobility (IM) method of separation. In modern instrumentation, IM typically follows an HPLC separation and is used in connection with a time of flight (TOF) mass spectrometric (MS) detection. While HPLC separation provides separations in the range of seconds, IM provides the separation of sample components in the range of 10-1 to 10-3 seconds, and TOF–MS produces the separation of ions by their mass on a microsecond scale. The IM separation significantly cleans up mass spectral data of co-eluting peaks not separated by the liquid chromatography and adds descriptive information for each ion. The combination LC–IM–MS allows separation based on the LC retention time, cross-sectional area differences in IM, as well as m/z and mass spectral fragmentation in the MS. The chapter describes several aspects regarding the types of IM and presents in detail the results of using LC–IM–MS in three case studies.

Each chapter is written by scientists with considerable experience in the field and recognized academic experience. This allows for the material to be clearly presented as well as very informative from a scientific point of view.

Dr. Serban C. Moldoveanu
R.J. Reynolds Tobacco Co.,
Winston-Salem, NC, United States

Victor David
Professor,
University of Bucharest,
Bucharest, Romania

Chapter 1

Progress in Technology of the Chromatographic Columns in HPLC

Serban C. Moldoveanu and Victor David

Abstract

Chromatographic column is an essential part of a any HPLC separation, and significant progress has been made in developing columns with better performance to provide better separation, a shorter separation time, resilience to a wider pH range of the mobile phase, longer lifetime, use of lower volumes of mobile phase, etc. All these characteristics were achieved by the introduction of novel technologies and improvements of the older ones. These include smaller particle used to fill the column, more homogeneous spherical particles, core-shell particles, monolithic columns, more pure silica as a stationary phase support, use of ethylene bridge silica, a wider variety of active phases, use of mixed mode stationary phases, use of polymers as stationary phase, use of various endcapping techniques, etc. Miniaturization and progress in the instrumentation played an important role for the chromatographic column development. All these aspects are summarized in the present chapter.

Keywords: chromatographic column, silica, support derivatization, reversed phase, HILIC, ion exchange, chiral columns

1. Introduction

The chromatographic separation is based on the differences in the retention of the components of a sample dissolved in a mobile phase when passing through a stationary phase typically contained in a chromatographic column. In HPLC, the mobile phase is a liquid and the characteristics of high performance (of the separation) and high pressure (used for the mobile phase) lead to the acronym HPLC. Although cartridges and micro-fluidic chips can be used to contain the stationary phase, a column is much more frequently utilized for this purpose [1, 2]. The external body of the column is a tube made from stainless steel or a strong polymer (e.g., polyether ether ketone or PEEK). This tube is filled with the stationary phase. Stationary phase can be in the form of particles or as monoliths. Both particles and the monoliths usually have a rigid porous support that may also act as the active phase, but more frequently the support has on the surface a chemically bonded or physically coated active phase used for the separation. The progress in the making of chromatographic columns is very important for the development of HPLC. A large body of information describes the progress in column construction including peer reviewed papers, books, and information on the Internet [3–11]. Present chapter describes some of the more recent progress in column construction and indicates potential for new developments. This progress takes place into two main directions: 1) the improvement of mainstream-type columns that are widely used in everyday work for practical analyses and 2) the development of exploratory new materials for the stationary phase and miniaturization.

2. Short theoretical background

In a chromatographic separation, the components of a mixture are eluting from the column, then are detected, and the detection electric signal is converted into a graphic output as *peaks* in a chromatogram. The peaks have ideally a Gaussian shape. Each peak has a specific *retention time* t_R. For a compound X, $t_R(X)$ is the time (usually measured in min) from the injection of the sample into the chromatographic system to the time of elution of the compound. A time slightly longer than the retention time of the last peak in a chromatogram is indicated as *run time*. The retention time for an unretained compound is known as *dead time* t_0. The dead time t_0 is defined by the ratio L/u where u is the *linear flow rate* of the mobile phase and L is the *column length*. In HPLC instrumentation, the controlled parameter by the user is the *volumetric flow rate* U and not u. The two parameters, U and u, are related by the expression:

$$U = \frac{\varepsilon^* \pi d^2}{4} u \tag{1}$$

In formula (1), d is the inner diameter of the column, and ε^* is a constant depending on column packing porosity (an average value for ε^* is 0.7 although this may vary considerably depending on the stationary phase particle size and structure).

The separation in a chromatographic process between two compounds X and Y (X eluting first in the separation) is overall characterized by a parameter termed *resolution* R. The expression for R is given by the formula:

$$R = \frac{1}{4}(\alpha - 1)\frac{k'(Y)}{1 + k'(Y)} N^{1/2} \tag{2}$$

In formula (2) parameter k' termed *retention factor* is defined by the formula:

$$k'(X) = \frac{t_R(X) - t_0}{t_0} \tag{3}$$

Parameter k' depends on chemical nature of the separated compound X, on the nature of the mobile phase, on the chemical composition and physical characteristics of the stationary phase as well as on a parameter termed *phase ratio* Ψ. The value of Ψ is given by the ratio V_{st}/V_0 where V_{st} is the volume of the active part of the stationary phase involved in separation process, and V_0 is the dead volume of the column ($V_0 = t_0\, U$). The retention factor k' is proportional with Ψ. Parameter α is the *selectivity*, which is defined as the ratio $k'(Y)/k'(X)$, and N is a parameter termed *theoretical plate number*, which describes the peak broadening in a separation and estimate the column efficiency. The value of N depends on column length and a related parameter to N independent on L is the *height equivalent to a theoretical plate* H (HETP) defined as L/N.

For achieving a good separation, the value of R should be higher than 1.0 and R is larger when α, N and k' are larger. These parameters can be increased by improving the columns properties. Since a larger k' indicates a longer retention time t_R, which is not usually desired, the increasing of R is achieved mainly by increasing α and N. The increase of N for a given L is achieved by decreasing H. The value of H depends on the linear flow rate u by the following expression known as van Deemter equation:

$$H = A'd_p + B'\frac{D}{u} + C'\frac{d_p^2}{D}u \qquad (4)$$

In formula (4), D is the diffusion coefficient of the mobile phase, d_p is the diameter of the particles in the column, and A', B', and C' are coefficients that depend on the nature of stationary and mobile phase.

In addition to R, many other parameters are used for the characterization of an HPLC separation such as peak asymmetry As, which shows the deviation from the ideal Gaussian shape of the chromatographic peak, column backpressure Δp, which is the difference between the pressure at the column inlet and that at the outlet of the column, etc. Column backpressure is described by the following formula (known as Darcy equation):

$$\Delta p = \frac{\eta u \phi_r L}{d_p^2} = \frac{\eta \phi_r L^2}{d_p^2 t_0} \qquad (5)$$

In formula (5), η is the mobile phase viscosity, and ϕ_r is a column flow resistance factor. Also, the columns are characterized by several other parameters and properties such as the construction of particles (porous, core-shell, etc.), uniformity of particles dimensions, porosity of the stationary phase, percent coverage of the solid support with the active phase, resilience of the stationary phase to a specific pH range of the mobile phase, resilience to dewetting, etc. A more detailed description of many parameters and properties describing a separation and characterizing the chromatographic column can be found in various books about HPLC (e.g., [12]). Significant progress in HPLC is being made such that to obtain better separation (higher R), shorter run times, lower values for As and Δp, reproducible separations, etc. A main source of progress is the improvement in the making of the chromatographic column by modifying parameters such as L, d, d_p, ε^*, k', H, As, ϕ_r, etc., in a manner that will lead to better chromatography. Other properties of the modern columns that are not captured with these parameters are also being improved and will be further discussed.

3. Trends in the column physical dimensions

The column physical dimensions are its length and internal diameter (i.d.). The common column lengths are between 30 mm and 250 mm with typical lengths of 50, 100, 150, 250 mm. The i.d. of the column is used to classify the columns as standard (3.0–4.6 mm i.d.), minibore (2.0–3.0 mm i.d.), microbore (0.5–2.0 mm i.d.), capillary (0.2–0.5 mm i.d.), and nanoscale (0.05–0.2 mm i.d.). The tendency of modern columns is to have them shorter and narrower leading to shorter run times and the use of less solvent. However, for a given H, the decrease in column length L leads to a lower value for N. The improvements in the stationary phase such that the columns have lower H allow the use of shorter columns maintaining a desirable R.

The use of narrower columns leads to higher linear flow rate u for a given volumetric flow rate U resulting in shorter retention times. This can be seen based on Eq. (3) and from the dependence of t_0 on u that give $t_R = (k' + 1)L/u$. Although sorter retention times are desirable, the increase in u is limited by the decrease in the value of H (as indicated by van Deemter Eq. (4)) and by the increase in column backpressure (as indicated by formula (5)). For this reason, the most commonly utilized columns in current practice are those with standard and minibore i.d. The use of microbore, capillary, and nanoscale column encounters problems with a decrease efficiency (decreased α, increased H) [13, 14]. In addition to that, the microbore and narrower columns have a low loading capacity (maximum amount of sample that can be loaded on the column) leading to requirements for the increased sensitivity of the detector. For narrower columns, a compromise in setting U must be made such that a faster chromatography is obtained but the associated increase in the value of H does not preclude a good separation. A study for the evaluation of the possibilities to use narrower columns indicated that an optimum i.d. is around d = 1.5 mm, which achieves short retention times and low solvent use with good column performance [5].

The tendency to use shorter and narrower column in order to achieve shorter run times and use of less volume of mobile phase will continue in the future [15]. The production of columns with smaller H, higher α, and the progress in the instrumentation allowing the use of higher backpressure for the chromatographic columns as achieved currently with ultrahigh-pressure chromatographs (or ultra-performance LC, UPLC) that can generate up to 1300 bar, will continue to allow the decrease in column length and diameter. In parallel with the developments of commonly used columns for routine analytical laboratories, significant effort is made in developing novel experimental columns using miniaturization, very high column backpressure, as well as special active phases, etc. The progress in the sensitivity of the detectors used in HPLC/UPLC will allow the use of smaller and more diluted samples to overcome the lower loading capacity of smaller columns.

4. Trends in the structure and composition of solid support of stationary phase

The most common type of stationary phase in HPLC and UPLC is made from small particles (typically 1.7–10.0 μm in diameter), which are packed in the body of the column. Monolithic columns are also utilized and are made from a single rod of a solid porous material. Because hydrated porous silica can have a very large surface and can be derivatized to bind an active phase, it is the most common material used as solid support to make the particles and also some monoliths for HPLC. This silica usually has a bonded, grafted, or coated layer of organic material. This organic layer is the active part of stationary phase involved in the separation process, but the silanol groups from the uncovered surface of silica also participate in the separation. In case of hydrophilic interaction liquid chromatography (HILIC) and in normal phase chromatography (NPC), bare silica can be used as active phase without additional coverage due to its polar character.

Not only silica can be used as support for the active stationary phase. Materials such as hybrid organic–inorganic still based on a hydrated silica but containing organic groups such as -CH_2-CH_2- in its structure can be used as support. Also hydrated zirconia, titania, ceramic hydroxyapatite, or organic polymers can be used as solid support. The progress regarding the solid support is made in two directions, one being the physical characteristics of the support and the other its chemical properties. These characteristics are further discussed separately.

4.1 Physical characteristics of stationary phase support

For particles used as solid support, one first characteristic is the physical type, which can be fully porous, core-shell, or pellicular. Porous particles (1.7–10 μm in diameter) have a porous structure for the entire particle, core-shell have a solid nonporous core 1.5–3 μm in diameter surrounded by a porous outer shell 0.3–0.5 μm in depth. Pellicular particles are solid nonporous spheres covered with a thin layer of stationary phase. Fully porous and core-shell particles are widely utilized in common HPLC practice, while pellicular particles are less common because of their reduced loading capacity. Core-shell particles offer better peak shape (lower H values) compared with fully porous particles and are likely to continue to be used even more frequently in the future. Generally, they are characterized by higher values for phase volume ratio Ψ than monolithic columns, but lower Ψ than fully porous particles [16].

The particles are also characterized by (average) diameter d_p, the shape of the particles, which can be irregular or spherical, the uniformity of the particles dimension, the surface area, the pore size and volume, the tortuosity and the uniformity of the channels in the particle, the structural rigidity. The dimension of particles d_p with diameters of 5 μm, 3 μm, 2.1 μm, 1.8 μm, 1.7 μm is commonly used, and smaller particles lead to lower H values as indicated by formula (4). An empirical formula shows how N depends on d_p as follows:

$$N \approx \frac{1000\,L}{Ct \cdot d_p} \tag{6}$$

In formula (6), Ct can be 2, 2.5, or even 3, depending on other particle characteristics. The use of core-shell particles and dimensions of 1.7–1.8 μm leads to columns having the values for N per m (L = 1 m) as high as 200,000–300,000 [17].

At the same time with increasing N, lower d_p leads to higher Δp as indicated by formula (5). This increase imposes limitations on how small the particles can be made. However, the tendency to use smaller particles to obtain columns with higher N is likely to continue in the future in spite of the increase in the backpressure, as the capability of the HPLC/UPLC pumps to deliver higher pressures increases.

Regarding particle shape, spherical particles show lower H values compared with particles of irregular form, and more uniform values of particle size show lower H values compared with particles of various sizes (particle size distribution is described by a parameter d_{90}/d_{10} and values lower than 1.2–1.3 indicate good homogeneity). Particles in modern columns have spherical shape and low d_{90}/d_{10} values, and these characteristics will continue to be maintained. Similar to the external aspect of particles, the internal structure regarding the channels in the porous material can be more or less homogeneous. The uniformity of particles interior is also contributing to a lower H value for the column.

Another physical characteristic of the stationary phase is its surface area [18]. For silica, common values for surface area are between 100 m^2/g and 300 m^2/g. The trend for modern stationary phases is to have particles with larger surface area since they can be coated with more active phase (increased Ψ value). However, the strength of stationary phase tends to decrease when stationary phase surface area increases. High silica strength (HSS) support is available now, allowing its use without restrictions in UPLC-type conditions where the back pressure of the column can be up to 900–1000 bar.

The pore size of the porous solid support is commonly characterized as small pores (with diameter below 60 Å), medium (in the range 60–150 Å), and large (of about 300 Å or larger). Common silica pore size is around 100 Å. However, the selection of the pose size depends on the type of molecules to be separated on the stationary phase. For small molecules (with *Mw* lower than about 3000 Da), the pore size of 100 Å is adequate, but for larger molecules, special phases with large pore size (about 250 Å) should be used. The adequacy of the pore size for the type of molecules to be separated, and in particular for the separation of proteins, is a field where significant development takes place [19].

Monoliths have a porous structure characterized by mesopores (pores between 2 and 50 nm in diameter) and macropores (about 4000–20,000 nm in diameter). For silica monoliths, the silica skeleton is 1–2 μm thick and has a void volume of almost 80% of the entire column volume. Polymeric monoliths have similar void volume. Since monolithic columns produce a lower pressure drop as compared with columns containing particles with similar characteristics, the monoliths are a promising material to be used as support for chromatographic columns. Monoliths are also successfully utilized in the construction of capillary and nano-LC columns [20].

4.2 Chemical characteristics of stationary phase support

The two main aspects of the chemical characteristics of solid support, which are of interest, include: 1) its internal chemical composition and 2) the chemical functionalities allowing the binding of the active phase (in cases when the solid support does not act itself as the active phase). Regarding the internal chemical composition, the solid support can be made from silica, ethylene or propylene bridged silica, hydrated zirconia, hydrated alumina, aluminosilicates, porous graphitic carbon, zeolites, or various organic polymers such as polystyrene cross-linked with divinylbenzene (PS-DVB) [21], methacrylates, etc. More recently, metal–organic frameworks (MOFs) were experimentally evaluated as support for HPLC stationary phases [22].

The most common support material is hydrated silica, which is obtained in principle from a chemical reaction that generates silicic acids followed by condensation reaction of the type:

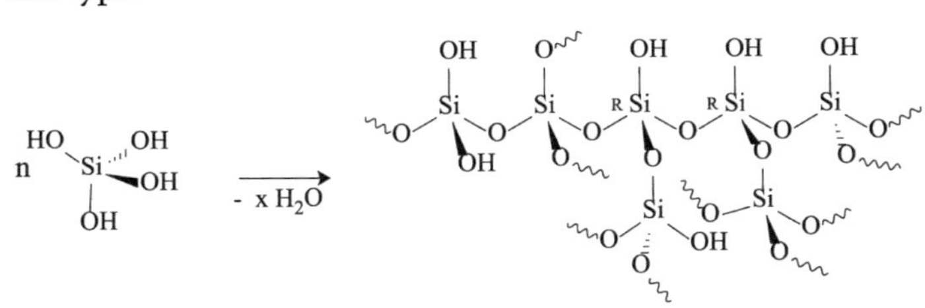

(7)

The resulting material contains numerous silanol groups that are further used for bonding the active phase. The purity of resulting hydrated silica is very important since the presence of metal ions in its structure leads to undesired effects such as peak tailing in chromatography. Very high purity silica (indicated as *Type B*) is now common as support in chromatographic columns. The ethylene bridge silica (indicated as BEH technology by Waters or TWIN technology or EVO by Phenomenex) [23] is also a common support offering excellent resilience to the strong acidic or basic character of the mobile phase (pH range of stability 1 to 12). This is a significant advantage compared with the range of pH stability of common silica, which is between 2.5 and 7.5. Ethylene bridge silica can be prepared from hydrolytic condensation of bis (triethoxysilyl)ethane and tetraethoxysilane using a small amount of water in a reaction schematically written as follows:

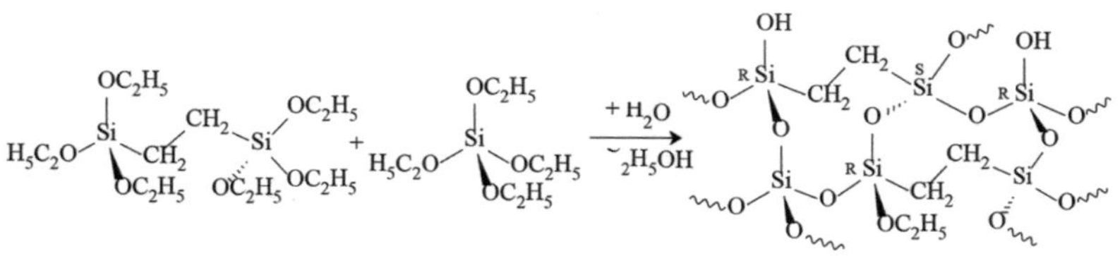

(8)

Ethylene bridge silica is an excellent material to be used as solid support for the stationary phase in HPLC, and its use will continue probably becoming even more common.

Regarding the other materials, columns based on hydrated zirconia are commercially available, but in general, they have lower chromatographic performance compared with those based on silica mainly due to numerous Lewis acid sites present on the stationary phase. Commercially available are also porous graphitic carbon columns. In order to achieve a large surface area, graphitic stationary phases are made using silica as template on which a layer of an organic material is applied followed by pyrolysis in an inert atmosphere to generate graphite. This is followed by the dissolution of silica template [24]. This type of column has a strong hydrophobic character, but some problems with surface homogeneity remain to be solved.

For the organic polymers, various procedures are used to obtain porous materials [21]. In some cases, these porous polymers may also act as the active phase, and in other cases they contain reactive groups on which an active phase is further bound. Among the problems with organic polymers as support material are the limitation of their structural rigidity and propensity to swelling in certain mobile phase compositions. Although for some types of HPLC (e.g., reversed phase or HILIC), the use of silica-based support is by far more common, organic polymers are frequently used in ion exchange chromatography and in size exclusion chromatography. Also, organic polymers are frequently used for making monolith-type columns [25].

The second important property of the stationary phase support is its capacity to react with a derivatization reagent with the goal of attaching a desired type of functionality such as aliphatic chains of 8 or 18 carbon atoms (C8 or C18) typically used in reversed phase (RP) type of HPLC. For silica support, the reacting capacity is assured by the presence of numerous silanol groups on the silica surface. The number of OH groups per unit mass of silica is characterized by silanol density α_{OH} expressed by the formula:

$$\alpha_{OH} = 602.214 \frac{\delta_{OH}}{S_{surf}} \tag{9}$$

In formula (9), S_{surf} is the surface in m^2/g, and δ_{OH} is the amount of silanol groups (in mmol/g). The value of α_{OH} typically varies between 4.1 and 5.6 OH groups per nm^2.

A different type of phase support still based on silica but with different active groups is hydride-based silica (known as type C silica [26]). This material is obtained using a reaction of the type indicated below:

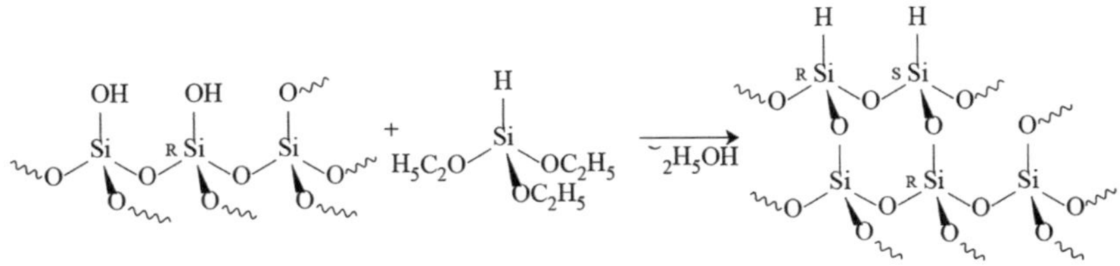
(10)

This type of silica can be used as normal phase without further derivatization or can react to attaching further organic groups that will operate as active phase.

Reactive groups used for further derivatization can also be present in various organic polymers. For example, many acrylate-type polymers are synthesized to contain glycidyl groups. These act as reactive sites on the porous polymer surface on which the desired functionalities can be bound. The use of organic polymers as solid support is an attractive alternative in particular related to the efforts toward miniaturization of HPLC columns, where 3D printing technology can be applied to make capillary and nano columns [27].

As described in this section, the most common stationary phase support is based on silica. Although numerous other types of support are continuously evaluated, significant progress is also being made in generating silica with better properties. One of the most promising directions is the preparation of ethylene-bridge silica, which offers an excellent stationary phase support, with good reactivity for binding the desired functional groups and with high resilience to the mobile-phase extreme pH values or composition. The use of ethylene-bridge silica in combination with core-shell type phase will continue to expand, and further progress is likely to continue for this type of phases.

5. Trends in the making of the active part of stationary phase

In many types of chromatographic columns, the active phase intended to be involved in the separation is bound or coated on a porous solid support with a large surface area. In some types of HPLC/UPLC, the solid support acts as the active phase without being further chemically modified, and some details about this type of phases will also be further presented, but this section is dedicated to bonded active phases. The bonded phase on a solid support is a key part of the type of chromatography for which the phase is made. For example, for RP-HPLC, which is the most common type of chromatography, the bonded phase is made to have a hydrophobic character. For this purpose, hydrocarbon moieties with different number of carbon atoms are attached to the porous support. The most common such groups contain 18 aliphatic linear carbon chains (C18) or eight carbons chains (C8), but other hydrophobic groups can be bound. For HILIC-type columns, organic fragments containing diol groups, amide, amino, sulfonylethyl, etc., can be bound. For ion-exchange-type chromatography, the bonded groups can be -COO$^-$, -SO_3^-, or -NH_3^+,- $N(CH_3)_3^+$, etc.

5.1 Progress in chemical reactions used for generating the active phase

Various chemical reactions are utilized for derivatizing the solid porous support of a stationary phase. The active phase can be directly attached to the silica surface, but variants of this procedure including the use of a pre-derivatization followed by a second one are also used. A typical derivatization reaction can be written as follows:

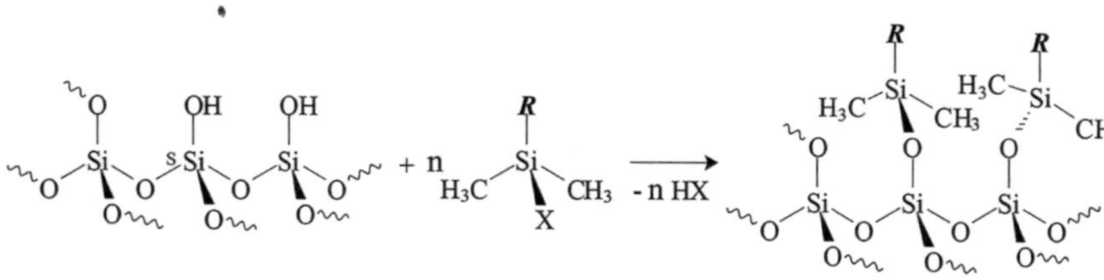

(11)

The reactive substituent X can be Cl, but also OCH_3, OC_2H_5, etc. The substituent ***R*** will determine the active phase (C8, C18, amino, cyano, and many others). Numerous variants of reaction (11) were applied for the derivatization of the silica solid support. In some of these variants, the CH_3 groups are replaced with other reactive substituents such as OC_2H_5, and the resulting material has the capability to further react. The procedure of using di- or tri-functional reagents (containing two or three reacting groups) leads to surfaces with different degree of coverage [28]. One important type of variant in derivatization is the formation of a single layer of attached active groups (indicated as *horizontal polymerization*) [29], or the formation of a multiple layer polymer on the silica surface (*vertical polymerization*). In vertical polymerization, a small amount of water is usually added during the derivatization process such that some of the bounded groups containing reactive fragments such as OC_2H_5 will be hydrolyzed generation active -OH functionalities that can be further reacting with the derivatization reagent. In this manner, the derivatization can be repeated a number of times [30]. For the preparation of hydrophobic phases with the use of vertical polymerization, various levels of carbon load (C%) can be placed on silica surface, C% varying depending on the procedure between 5% and 30%.

As the sensitivity of detection in HPLC/UPLC is becoming higher and higher in particular with the development of mass spectrometric (MS) and MS/MS detectors, one important quality of the chromatographic columns is to have a very low background that may be caused by small molecule "leaking" into the mobile phase left from the manufacturing process of the stationary phase. The use of trifunctional reagents and new procedures to achieve the derivatization of the solid support (usually of silica) led to chromatographic columns with very low bleed, higher resilience to a wide range of composition for the mobile phase and good reproducibility of the separation.

After derivatization, silica surface (and also the surface of hydrated zirconia or alumina) still remains with a considerable number of underivatized OH groups (silanols in case of silica). These silanol groups interact with the analytes from the mobile phase such that not only the ***R*** groups forming the active phase influence the separation but also the silanols. This effect is undesirable in some cases, and the process of endcapping is used for diminishing (or removing) the silanol interference. The endcapping consists of additional derivatization that places on silanols small organic groups such as $Si(CH_3)_3$. Steric hindrance that precludes the dense covering of the silica surface with larger groups such as C8 or C18 is avoided by using derivatization with small groups such as trimethylsilyl (TMS). Repeated derivatization with the endcapping reagent such as chlorotrimethylsilane is usually performed when most silanol groups are intended to be covered.

The process of silica surface derivatization offers numerous possibilities to generate stationary phases with different properties [31]. A large variety of columns is commercially available, and they are tailored for specific utilization. Derivatization and endcapping are used to obtain stationary phases with higher resistance to extreme pH of the mobile phase (e.g., controlled surface charge or CSH type columns), with extra dense bonding (XDB) of the active phase, with different degrees of hydrophobicity, with polar endcapping groups (e.g., CH_2OH), or with embedded polar groups in the hydrophobic chain of the active phase [32, 33]. Besides the procedures summarily indicated above to derivatize the solid support, various other derivatization procedures are reported in the literature [34]. Also, alternative procedures to obtain the active phase such as direct synthesis of silica materials with an active bonded phase surface [35] can be used. Progress in the synthesis of monoliths, as well as of stationary phases based on organic polymers, is also being made [36]. One such example is the production of latex-agglomerated ion exchangers.

A variety of other procedures are available for producing the active phase for HPLC and UPLC columns (e.g., [37]). Some of these procedures are kept undisclosed by the column manufacturers and some are reported in the literature. Also, a variety of novel procedures for attaching the active phase on the solid support are developed, such as grafting of pre-synthesized polymers [38, 39], or direct synthesis of the stationary phase containing the desired functionalities [40, 41].

Stationary phases with better performance including better resolution R, lower asymmetry As, resilience to a wider pH range of the mobile phase, capability to work in 100% aqueous mobile phase (resilience to de-wetting), and production of phases with more complex structure than a single functionality are achieved using a diversity of procedures to derivatize the porous solid support. The use of trifunctional derivatization reagents and special endcapping techniques was among the important procedures to achieve this goal, and the use of these procedures is likely to continue to be improved in the future.

5.2 Improved properties of active stationary phase

The modern columns have various benefits from the improvements in the synthesis of the active phase. For example, from the derivatization with trifunctional reagents, the active phase is more homogeneous and stable, with reduced access of the analytes to the free silanols and more reproducible chromatography. The horizontal polymerization (derivatization) has the advantage of higher homogeneity and reduced presence of free silanols, while vertical polymerization leads to phases with a higher mass of active phase (larger Ψ). The new active phases allow the separation to be based on wider types of interaction, and preparation of phases with mixed mode functionalities is more and more common. The progress made in the endcapping process, the capability to use polar endcapping, and the introduction of controlled surface charge (CSH) contribute to the extension of pH range of column stability. Also, the stability of columns in time (to be used for a larger number of injections), the low bleed allowing the use of the columns with very sensitive detectors without generating a high background signal are important factors in the increase of column quality.

Besides the making of columns with improved characteristics, the increased variety of available columns is another direction in which considerable progress is being made. This variety of columns allows a better selection for a specific task, and also, as bidimensional HPLC is sometimes needed for the separation of complex samples, the column variety offers choices for orthogonal separations [42].

6. Diversity of HPLC columns

Under the acronyms HPLC or UPLC are included a number of similar techniques that have significant differences regarding the mechanism involved in the separation. According to the separation mechanism, a specific type of chromatographic column is used. Some of HPLC techniques are common, and some are more special having lower utilization. One main type of common HPLC/UPLC is RP-HPLC, which is used for the separation of molecules having in their structure hydrophobic moieties but frequently additional polar groups. Other common HPLC types are HILIC used for the separation of strongly polar molecules, ion exchange HPLC used for the separation of molecules capable to ionize, chiral HPLC used for the separation of enantiomeric molecules, size exclusion HPLC used for the separation of molecules based on their molecular size (more precisely hydrodynamic volume), and affinity/immunoaffinity HPLC. Various other techniques less frequently utilized are derived from the main types, and examples of such techniques are ion pair chromatography, hydrophobic interaction, normal phase, ion moderated, etc. The active stationary phase for each of those techniques has specific structures. Regardless of the column type, all modern columns benefit from the progress in the solid support in particular by using high-purity silica and ethylene bridge silica, from the use of core-shell particle construction and the advances in the making of monoliths. Some specific aspects for different types of HPLC/UPLC are further discussed.

6.1 Columns for RP-HPLC

Frequently used for the analysis of a large range of compounds, from small molecules to proteins and from highly hydrophobic to rather polar ones, RP-HPLC is the most commonly applied HPLC technique. To this extensive use is associated a significant number of RP type columns many of them commercially available. For RP-HPLC the active stationary phase contains hydrophobic groups, the most common being C18 and C8 phases. The hydrophobic character of the stationary phase in RP-HPLC can be modified by using the active phase with specific groups. Besides C18 and C8 that are very common, aliphatic C2, C4, C12, C14, C20, C22, C27, C30, cyclohexyl, phenyl, diphenyl, C6-linked phenyl, pentafluorophenyl, cyanopropyl, etc., can be used to create a hydrophobic surface. The hydrophobic character of these phases represents one criterion to differentiate them. However, even for columns containing the same type of phase, such as C18, many variations in the active phase structure are possible. The variations may include the type of bonding (mono, di, or trifunctional), the type of polymerization (horizontal or vertical), the carbon load, the density and uniformity of the coverage of solid support (e.g. of silica), and the variations in endcapping. Some hydrophobic stationary phase may contain polar imbedded groups [43]. Various imbedded groups in aliphatic chains were reported in the literature [32], and some are present in commercially available columns. Some of these groups include ether, amide, urea, carbamate, sulfone, thiocarbamate, etc. These groups are used to modulate separation of many types of organic compounds that have in their molecule polar groups. In addition, the imbedded polar groups "shield" the silica residual silanols for interacting with the analytes (in particular with highly basic ones) leading to a reduced silanol activity of the stationary phase (as in Symmetry Shield type columns) and also to better resilience to extreme pH values of the mobile phase.

Evolution of stationary phases in RP-HPLC (and RP-UPLC) took place in two directions: 1) perfecting common columns such as C18 or C8 columns and 2) exploring the binding of various less common groups on the solid support. Perfection of common columns is being done by working with either fully porous or core-shell particles, using special substrates usually high-purity silica or ethylene bridged silica, controlling the derivatization to be very homogeneous, and using special endcapping. By endcapping with TMS groups, the polarity of the silanols is reduced, but the extent of this process can vary from column type to column type, and some C18 columns are intentionally left with some silanol activity for interacting with polar molecules. The use of endcapping with small polar groups also brings distinctive properties to the RP-type columns. Adding special procedures such as CSH or XDB technologies, the variety of RP columns becomes even larger. CSH technology takes advantage that the silica surface is usually slightly negatively charged due to the dissociation of silanols. This charge can be neutralized by adding specific reagents such that the surface reactivity is decreased. The technology is applied to ethylene bridge particles by incorporating a low level of surface charges on stationary phase particles. Also, the construction of phases with C18 or C8 active phase but based on silica with specifically larger pores (e.g., 250 Å) is a promising path for the separation of large molecules such as proteins.

Regarding the binding of various less common groups on the solid support, special phases with bonded cholesterol or fullerene moieties were made, as well as columns with aliphatic chains having an unusual number of carbons (e.g., C3 or C4 for lower hydrophobicity or C30 for intended higher hydrophobicity) [44]. However, these types of experimental bonded phases did not generate columns with much different hydrophobic properties. The intimate mechanism of hydrophobic interactions caused by the "rejection" of the molecules containing hydrophobic moieties from a polar solvent and their "acceptance" in a hydrophobic stationary phase leads to a non-unique process of separation, as long as the accepting phase is less polar than the mobile phase (e.g., [12]). As a result, the choice of mobile phase composition in RP-HPLC plays an important role in the separation, and the differences in the properties of columns used in RP-HPLC are basically obtained by modulating the ratio of hydrophobicity and residual polar interactions and less by changing the phase hydrophobicity.

The use of hydrated zirconia as solid support, the use of coating of a silica base and not binding it, the use of organic polymers to make phases for RP-HPLC, or the use of porous graphitic carbon as stationary phase, although leading to a variety of columns to be used in RP-HPLC remained with a relatively limited utilization. Both trends of improving columns with common stationary phase such as C18, C8, phenyl, cyanopropyl, and experimenting with new active phases are likely to continue in the future. However, a considerably more impact for the progress is still expected from the improvements of common stationary phases.

6.2 Columns for HILIC and NPC

Important progress has been made in the construction of columns dedicated to HILIC separation. The active phase for these columns must be polar, and it is used with a mobile phase less polar than the stationary phase and containing water plus an organic solvent. Similar phases are used for NPC, but in this case the mobile phase is non-aqueous. Bare silica can be used as stationary phase in HILIC, and the improvements in the silica purity and homogeneity of silanol coverage made these columns rather common. Bonded phases with groups such as diol, ether embedded+diol, amide terminal, polyamide, cyano (also used in RP HPLC) are common. Propylamine, diethylamine, or triazole groups are used to generate weak anionic active phases, sulfonylethyl groups are used to generate weak cationic active phase, and amino +sulfonic, amino+carboxylic groups are used to generate zwitterionic phases. Various other types of phases for HILIC applications were synthesized [45]. These phases have various polarities, but the spacer (handle) molecular fragment connecting the polar group with the silica base plays an important role in the separation. The same features as for RP-HPLC columns, including the coverage of support with the bonded phase, the pH resilience, the preparation procedure using mono-, bi-, or tri-functional reagents, the phase ratio are important for the column quality. Since in the HILIC separations not only the polar interactions are important in the separation, but also the hydrophobic interactions play a role, the carbon load (caused by the spacer) also influences the separation characteristics. Some HILIC columns are also endcapped, and this process changes the stationary phase characteristics in a similar manner as for the RP-HPLC. Besides common phases used in HILIC separations, special stationary phases were also known. Such phases were made with bonded cyclodextrin, bonded perhydroxyl-cucurbit[6]uril, polyhydroxyethyl-aspatamide, polysuccinimide [46], etc. One example of a structure of a zwitterionic stationary phase containing sulfonylalkylbetaine groups used in HILIC separations is indicated below:

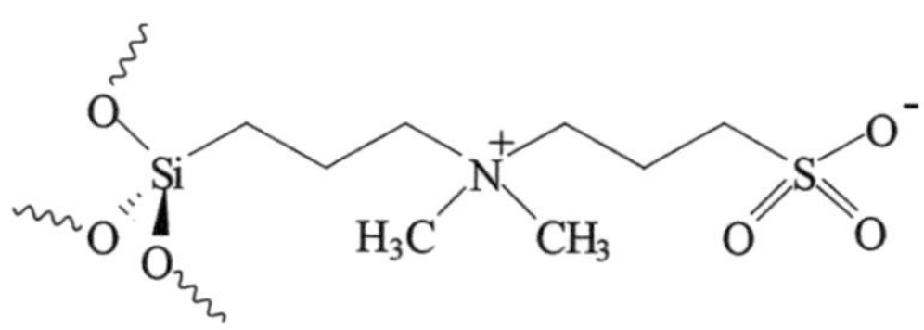

(12)

Because of the proximity of the positive and negative charged groups in the structure, the phase is not used as a zwitterionic ion exchanger.

Stationary phases based on organic polymers are also used for HILIC separations [47]. However, more common are still the silica-based columns.

6.3 Columns for ion exchange HPLC and related techniques

Ion exchange (IC) stationary phases are classified as cation exchange phases (weak, medium, and strong), anion exchange phases (also weak, medium, and strong), zwitterionic, and amphoteric. The phases contain groups attached through a handle on silica or on an organic polymeric support. Specific groups such as $-COO^-$, $-PO_3H^-$, $-SO_3^-$, etc., generate cationic phases, groups such as $-NH_3^+$, $-NH_2(CH_3)]^+$, $-N(CH_3)_3]^+$, $-[N(CH_3)_2(CH_2CH_2OH)]^+$, $-[N(C_2H_5)(CH_3)_2]^+$ generate anionic phases, and groups such as $-N(CH_3)_2^+-(CH_2)_n-SO_3^-$ or $-CH(SO_3^-)-(CH_2)_n-N(CH_3)_3^+$ generate zwitterionic phases. While for RP-HPLC and HILIC phases, the use of organic polymeric support is less common, for ion exchange phases the use of polymeric support is more common. A specific type of polymeric support is the latex agglomerated type. The latex agglomerated ion exchange particles contain an internal core that has ionic groups on its surface. On this surface is attached a monolayer of small diameter particles that carry functional groups having bonded ions with an opposite charge with those of the support. The groups of the outer particles have the double role of attaching the small particles to the support and also to act as an ion exchanger for the ions in the mobile phase. The advantages of this type of phase include its stability to a wide range of pH of mobile phase and resilience to higher column backpressure compared with common polymeric columns. This is possible because the cross-linking of the polymer from the core particles can be very high.

Because the loading capacity for the same amount of stationary phase is typically larger for IC columns compared with RP or HILIC columns, and because the separation mechanism is based on ionic interactions, which is different from that in RP-HPLC and HILIC, the capillary columns in IC are more successfully utilized. Such columns are used with a low flow rate (e.g., 0.01–0.02 mL/min) that increases the sensitivity of the conductivity detector used frequently in IC separations [48, 49].

Ion chromatography is extensively used in the separations of proteins and nucleic acids [50], and continuous progress is being made with new phases of IC type. Many such new phases are commercially available [10].

Special ion chromatographic columns are also applied in ion-moderated and ligand exchange chromatography. These types of columns are used for the separation of carbohydrates, sugar acids, as well as lipids. For example, difficult separation such as those between cis and trans lipids and fatty acids can be achieved using an ion-moderated columns containing Ag^+ ions [51]. In spite of the need for ion-moderated chromatography for the separation of important types of analytes, some of the existent columns dedicated for ion-moderated chromatography require relatively long run times for the separation. For this reason, development of new ion-moderated type columns would be highly desirable.

The ion exchange stationary phases and columns are in continuous development, and in particular mixed mode phases containing ion exchange type moieties are demonstrated to be very useful in separations. A discussion dedicated to mixed mode phases is also included in this chapter.

6.4 Columns for chiral separations

The increased demand of analysis of a variety of pharmaceutical drugs, many of them with chiral character, required constant development of chiral columns. Other fields of chemical analysis also required chiral separation. For example, the increased use of vaping and the proliferation of companies producing synthetic nicotine required the development of sensitive methods for the analysis of nicotine enantiomers [52]. Active stationary phase for chiral separation can be of different types, which include: brush or "Pirkle" type, cellulose based, cyclodextrin or cyclofructan-based, amylose-based, crown-ether-based, macrocyclic antibiotic type, protein based, ligand exchange type, chiral synthetic polymer type [53, 54], etc. All these phases contain various types of chiral centers. In spite of the existence of such a variety of columns, the need for stationary phases offering better enantioresolution is still actual. Many chiral columns must be used in non-aqueous mobile phase (NPC type chromatography), and fewer phases allow the use of water in the mobile phase for RP, HILIC, or IC-type utilization. However, many chiral compounds are highly polar and some are even insoluble in non-aqueous media. In addition, the widespread electrospray type of MS detection (ESI-MS) generates weak or no response when a mobile phase with no water is used for the separation. For these reasons, continuous effort is made to develop chiral columns that work in RP, HILIC, or IC mode.

The improvements of stationary phases for chiral separations follow the same lines as the one utilized for other types of columns. The use of core-shell type particles (e.g., [55]), smaller particle size, monoliths, various types of phases containing chiral centers such as glicopeptides, and macrocyclic antibiotics, as well as more common ones such as derivatized polysaccharides [56, 57] is providing important tools for obtaining better, more efficient types of chiral chromatographic columns [58].

6.5 Columns for size exclusion HPLC

Size exclusion HPLC (SEC) is a technique used for the separation of analytes according to their molecular size (hydrodynamic volume), and it is applied for the separation of macromolecules of different sizes and of macromolecules from small molecules. Ideally, only the size of the molecule should contribute to the separation, but it is common that some energetic interactions (e.g., of polar type) also take place between the stationary phase and the analytes. These energetic effects can modify the intended purpose in which only the size affects the separation. As the molecular size is usually proportional with the molecular weight *Mw* of a molecule, size exclusion is also used for the evaluation of *Mw* for macromolecules. Depending on the solubility of the polymers in an aqueous solvent or in an organic solvent, SEC is indicated as gel filtration chromatography (GFC) or as gel permeation chromatography (GPC), respectively. The stationary phases in SEC can be based on porous silica or on other inorganic materials, but very commonly on organic polymeric materials. Polymeric materials for making the stationary phase are more frequently used in SEC than in other HPLC procedures [59]. A common material for SEC stationary phase is polystyrene-divinylbenzene (PS-DVB) with different cross-linking degrees, but also gels based on dextran or agarose, hydroxylated poly(methyl methacrylate) (HPMMA), and polyvinylalcohol (PVA) copolymers are used. The separation phase should be made with large and controlled pore dimensions.

Among the requirements for a good stationary phase in SEC is to have homogeneous pores, to be as inert as possible and have minimal energetic interactions with the analytes, and to be resilient to high HPLC-type backpressure. These requirements are not very simple to achieve. The control of pose size such that they are as uniform as possible can pose difficulties during manufacturing. Silica-based SEC columns can be made using bare silica, but also bonded phases containing, for example, diol groups on silica are produced. The use of silica with large pores leads to lower resilience to the backpressure. In addition, the reduction of energetic interactions with the silanol groups on silica is not simple. For the polymeric phases, the problem of resilience to higher backpressure is even more stringent than it is for the silica-based phases. The use of special cross-linked polymers alleviates this problem. Also, SEC columns usually require long run times for separations, but new developments such as making core-shell type stationary phases shorten the separation time. Also, as the pressure resistance of the used materials is better, the reduction in the particle size of the phase contributes to improvements in SEC chromatography [60]. New stationary phases use all those procedures to improve the chromatographic columns for SEC.

6.6 Columns in affinity, immunoaffinity, and aptamer-type HPLC

In affinity/immunoaffinity chromatography (IAC), the stationary phase contains on its surface an immobilized biological complement of the analytes from the mobile phase [61]. Examples of pairs of biological complement and the analytes are antigens and their antibody, lectins and glycoproteins, metal ions and proteins containing amino acid residues that have affinity for the ion (e.g. histidine), biotin and avidin, etc. The solid support for the stationary phase can be silica, synthetic organic polymers, agarose (the neutral gelling fraction of the complex natural polysaccharide agar), cross-linked agarose, cross-linked dextrans (sepharose, sephacryl), cellulose, etc. It is typical for the solid support in affinity chromatography to have large pores, between 300 Å and 500 Å because the technique is used for the separation of large molecules (e.g., proteins and nucleic acids). The stationary phase particles can be porous] or nonporous [62] and also can be monolithic [63]. A variety of techniques are used to make stationary phases for IAC, using different procedures for the immobilization of biological complement ranging from covalent attachment to adsorption-based methods. For example, the immobilization of antibodies can be done through their amine groups by using a support that has been activated with reagents such as N, N′-carbonyldiimidazole, cyanogen bromide, N-hydroxysuccinimide, or tresyl chloride/tosyl chloride [64]. New phases are continuously reported for this technique, with a variety of active phases including different types of proteins, aptamers [65], and dye ligands [66]. Continuous progress is also made regarding stationary phases for biomimetic LC that mimic the interactions in natural biological systems [67, 68].

6.7 Mixed-mode HPLC columns

Preparation of stationary phases with mixed-mode active groups in which the separation is based on two or more types of main interactions is currently an important direction of development in HPLC [11]. Mixed-mode phases offer special separation capabilities and could be a simpler alternative to bidimensional separations that use orthogonal columns [69]. These phases may have reversed-phase and HILIC capabilities, reversed-phase and ion exchange capability, HILIC and ion exchange, or even more than two types of capability allowing for example reversed-phase/hydrophilic interaction/ion-exchange-type separations [70]. Some of the mixed mode phases also have chiral centers such that can be used for special chiral separations [71]. Porous or core-shell silica can be used for the preparation of mixed mode phases, and common functionalities such as C18, NH_2, diol, SO_3^-, etc., that are specific for one type of phase are used to obtain the mixed-mode phases. The main difference from single type of phase is that multiple functionalities are simultaneously present on the solid support. Synthesis of such phases frequently requires a sequence of derivatizations and strict control of the quality of the final product [72, 73]. The preparation of mixed-mode phases with organic polymers support, in the form of monoliths or using covalent organic frameworks, has also been described in the literature [74, 75]. Mixed-mode stationary phases can also be made as having the active functionality based on ionic liquids moieties [71, 76].

7. Conclusions

The chromatographic column is a key component of HPLC instrumentation, and the extensive use of HPLC promoted the effort for obtaining better columns. These columns provide better separations in a shorter time, generating reproducible chromatography, have minimal bleed avoiding background for the detectors, are resilient to a wide pH range of the mobile phase, can be used with the mobile phase having 100% water, and have a longer utilization life. Progress in making the chromatographic columns has been achieved by various procedures such as the optimizing the chromatographic column dimensions, the use of smaller particles for the stationary phase, the use of monoliths, the use of core-shell type particles. Significant progress was also made in chemistry of stationary phase, both regarding the solid support and the active phase bonded on it. Future progress is expected on the same lines of development for columns used in routine analyses. At the same time, experimental columns for HPLC miniaturization and enhanced efficiency are experimented and reported in the literature (e.g., [20, 77]). The parallel progress regarding the pumping system of HPLC instrumentation that can provide higher backpressure and well-controlled low flow rates, the precision of injecting systems (autosamplers), as well as the unprecedent increased sensitivity of detection in particular of MS and MS/MS type, were key for making possible some of the improvements in chromatographic column construction.

Author details

Serban C. Moldoveanu[1]* and Victor David[2]

1 R.J. Reynolds Tobacco Co., Winston-Salem, NC, USA

2 Faculty of Chemistry, Department of Analytical Chemistry, University of Bucharest, Bucharest, Romania

*Address all correspondence to: smoldov@aol.com

References

[1] https://www.waters.com/waters/en_US/ionKey-MS—microflow-UPLC-Seperation-with-iKey/nav.htm?cid=134782630&locale=en_US

[2] Bao B, Wang Z, Thushara D, Liyanage A, Gunawardena S, Yang Z, et al. Recent advances in microfluidics-based chromatography - a mini review. Separations. 2021;**8**:3

[3] Vissers JPC. Recent developments in microcolumn liquid chromatography. Journal of Chromatography A. 1999;**856**:117-143

[4] Recent developments in LC column technology, Supplement to LC/GC North America 36. 2018. No. s6

[5] Fekete S, Murisier A, Losacco GL, Lawhorn J, Godinho JM, Ritchie H, et al. Using 1.5 mm internal diameter columns for optimal compatibility with current liquid chromatographic systems. Journal of Chromatography A. 2021;**1650**:462258

[6] Bell DS. New chromatography columns and accessories for 2018. LC/GC North America. 2018;**36**:234-247

[7] Bell DS. New liquid chromatography columns and accessories. LC/GC North America. 2019;**37**:232-243

[8] Recent developments in LC column technology, Supplement to LC/GC North America 38. 2020. No. s6

[9] Lopez DA, Green A, Bell DS. What is on your HPLC particle? A look at stationary phase chemistry synthesis. LC/GC North America. 2020;**38**:488-493

[10] Bell DS. New chromatography columns and accessories for 2020. LC/GC North America. 2020;**38**:211-219

[11] Bell DS. Modern trends in mixed mode liquid chromatography columns. LC/GC North America. 2021;**39**:56-60

[12] Moldoveanu SC, David V. Essentials in Modern HPLC Separations. Amsterdam: Elsevier; 2013

[13] Lestremau F, Wu D, Szücs R. Evaluation of 1.0 mm i.d. column performances on ultra-high pressure liquid chromatography instrumentation. Journal of Chromatography A. 2010;**1217**:4925-4933

[14] Eggleston-Rangel R. Why use miniaturized columns in liquid chromatography? Benefits and challenges. The Column. 2021;**17**:21-25

[15] Blumberg LM. Column length is a structure-independent measure of solvent consumption in liquid chromatography. Journal of Chromatography A. 2022;**1662**:462727

[16] Caiali E, David V, Aboul-Enein HY, Moldoveanu SC. Evaluation of the phase ratio for three C18 high performance liquid chromatographic column. Journal of Chromatography A. 2016;**1435**:85-91

[17] Henry RA. Impact of particle size distribution on HPLC column performance. LCGC. 2014;**32**:12-19

[18] Giaquinto A, Liu Z, Bach A, Kazakevich Y. Surface area of reversed-phase HPLC columns. Analytical Chemistry. 2008;**80**:6358-6364

[19] Schuster SA, Wagner BM, Boyes BE, Kirkland JJ. Optimized superficially porous particles for protein separations. Journal of Chromatography A. 2013;**1315**:118-126

[20] Ghanem A, Marzouk AA, El-Adl SM, Fouad A. A polymer-based monolithic capillary column with polymyxin-B chiral selector for the enantioselective nano-high performance liquid chromatographic pharmaceutical analysis. Journal of Chromatography A. 2022;**1662**:462714

[21] Lloyd LL. Rigid macroporous copolymers as stationary phases in high-performance liquid chromatography. Journal of Chromatography. 1990;**544**: 201-217

[22] Xuan QQH, Zhang K, Chen X, Ding Y, Feng S, Xu Q. Core-shell silica particles with dendritic pore channels impregnated with zeolite imidazolate framework-8 for high performance liquid chromatography separation. Journal of Chromatography A. 2017; **1505**:63-68

[23] Wyndham KD, O'Gara JE, Walter TH, Glose KH, Lawrence NL, Alden BA, et al. Characterization and evaluation of C18 HPLC stationary phases based on ethyl-bridged hybrid organic/inorganic particles. Analytical Chemistry. 2003;**75**:6781-6788

[24] Pereira L. Porous graphitic carbon as a stationary phase in HPLC: Theory and applications. Journal of Liquid Chromatography Releted Technologies. 2008;**31**:1687-1731

[25] Nischang I, Teasdale I, Brüggemann O. Porous polymer monoliths for small molecules separations: Advancements and limitations. Analytical and Bioanalytical Chemistry. 2010;**400**:2289-2304

[26] Borges EM. Silica, hybrid silica, hydride silica and non-silica stationary phases for liquid chromatography. Journal of Chromatographic Science. 2015;**52**:580597

[27] Matheuse F, Vanmol K, Van Erps J, De Malsche W, Ottevaere H, Desmet G. On the potential use of two-photon polymerization to 3D print chromatographic packed bed supports. Journal of Chromatography A. 2022; **1663**:46273

[28] Hetem M, Van de Ven L, de Haan J, Cramers C. Study of the changes in mono-, di- and trifunctional octadecyl-modified packings for reversed-phase high-performance liquid chromatography with different eluent compositions. Journal of Chromatography. 1989;**479**:269-295

[29] Wirth MJ, Fatumbi HO. Horizontal polymerization of mixed trifunctional silane on silica. 2. Application to chromatographic silica gel. Analytical Chemistry. 1993;**65**:822-826

[30] Kirkland JJ. Development of some stationary phases for reversed-phase high-performance liquid chromatography. Journal of Chromatography A. 2004;**1060**:9-21

[31] Qui H, Liang X, Sun M, Jiang S. Development of silica-based stationary phases for high-performance liquid chromatography. Analytical and Bioanalytical Chemistry. 2011;**399**: 3307-3322

[32] O'Sullivan GP, Scully NM, Glennon JD. Polar-embedded and polar-endcapped stationary phases for LC. Analytical Letters. 2010;**43**:1609-1629

[33] Zhang Y, Chen M, Zhou S, Han H, Zhang M, Qiu H. A carbonylative coupling approach to alkyl stationary phases with variable embedded carbamate groups for high-performance liquid chromatography. Journal of Chromatography. A. 2022;**1661**: 462718

[34] Ashu-Arrah BA, Glennon JD, Albert K. Synthesis and characterisation of bonded mercaptopropyl silica intermediate stationary phases prepared using multifunctional alkoxysilanes in supercritical carbon dioxide as a reaction solvent. Journal of Chromatography. A. 2012;**1222**:38-45

[35] Hasegawa I. Co-hydrolysis products of tetraethoxysilane (TEOS) and methyltriethoxy-silane in the presence of tetramethylammonium ions. Journal of Sol-Gel Science and Technology. 1993;**1**: 57-63

[36] Oláh E, Fekete S, Fekete J, Ganzler K. Comparative study of new shell-type, sub-2μ fully porous and monolith stationary phases, focusing on mass-transfer resistance. Journal of Chromatography. A. 2010;**1217**: 3642-3653

[37] Moldoveanu SC, David V. Selection of the HPLC Method in Chemical Analysis. Amsterdam: Elsevier; 2017

[38] Yu S, Sha X, Zhou X, Guo D, Han B, Huang S, et al. Cyclodextrin-dendrimers nanocomposites functionalized high performance liquid chromatography stationary phase for efficient separation of aromatic compounds. Journal of Chromatography A. 2022;**1662**:462730

[39] Fan C, Quan K, Chen J, Qiu H. Comparison of chromatographic performance of co-grafted silica using octadecene respectively with vinylpyrrolidone, vinylimidazole and vinylpyridine. Journal of Chromatography A. 2022;**1661**:462690

[40] Wang X, Cheng S, Chan JCC. Propylsulfonic acid functionalized mesoporous silica, synthesized by in situ oxidation of thiol groups under template free conditions. Journal of Physical Chemistry C. 2007;**111**:2156-2164

[41] Zheng Y, Wan M, Zhou J, Dai X, Yang H, Xia Z, et al. One-pot method for the synthesis of β-cyclodextrin and covalent organic framework functionalized chiral stationary phase with mixed-mode retention mechanism. Journal of Chromatography A. 2022; **1662**:462731

[42] Pellett J, Lukulay P, Mao Y, Bowen W, Reed R, Ma M, et al. Orthogonal separations for reversed-phase liquid chromatography. Journal of Chromatography A. 2006;**1101**: 122-135

[43] Silva CR, Bachmann S, Schefer RR, Albert K, Jardim ICSF, Airoldi C. Preparation of a new C18 stationary phase containing embedded urea groups for use in high-performance liquid chromatography. Journal of Chromatography A. 2002;**948**:85-95

[44] Rimmer CA, Sander LC, Wise SA. Selectivity of long chain stationary phases in reversed phase liquid chromatography. Analytical and Bioanalytical Chemistry. 2005;**382**: 698-707

[45] Bo C, Li Y, Liu B, Jia Z, Dai X, Gong B. Grafting copolymer brushes on polyhedral oligomeric silsesquioxanes silsesquioxane-decorated silica stationary phase for hydrophilic interaction liquid chromatography. Journal of Chromatography A. 2021; **1659**:462627

[46] Alpert AJ. Cation-exchange high performance liquid chromatography of proteins on poly(aspartic acid) –silica. Journal of Chromatography A. 1983;**266**: 23-37

[47] Xu M, Peterson DS, Rohr T, Svec F, Fréchet JM. Polar polymeric stationary phases for normal phase HPLC based on monodisperse macroporous

poly(2,3-dihydroxypropyl methacrylate-co-ethylene dimethacrylate) beads. Analytical Chemistry. 2003;**75**:1011-1021

[48] Kuban P, Dasgupta PK. Capillary ion chromatography. Journal of Separation Science. 2004;**27**:1441-1457

[49] Wouters B, Bruggink C, Desmet G, Agroskin Y, Pohl CA, Eeltink S. Capillary ion chromatography at high pressure and temperature. Analytical Chemistry. 2012;**84**:7212-7217

[50] Mant CT, Hodges RS, editors. High-Performance Liquid Chromatography of Peptides and Proteins, Separation, Analysis, and Conformation. Boca Raton: CRC Press; 1991

[51] Momchilova S, Nikolova-Damyanova B. Chapter: Silver Ion Chromatography of Fatty Acids. Springer, Dordrecht: Encyclopedia of Lipidomics; 2016

[52] Moldoveanu SC. Interconversion of nicotine enantiomers during heating and implications for smoke from burn-down cigarettes, heat not burn devices, and vaping, Chirality. 2022;**34**:667-677

[53] Pei Y, Li X, Zeng G, Gao Y, Wen T. Chiral stationary phases based on lactide derivatives for high-performance liquid chromatography. Journal of Chromatography A. 2022;**1661**:462705

[54] Xie S-M, Yuan L-M. Recent development trends for chiral stationary phases based on chitosan derivatives, cyclofructan derivatives and chiral porous materials in high performance liquid chromatography. Journal of Separation Science. 2019;**42**:6-20

[55] http://www.azypusa.com/column-stationary-phases

[56] Younes AA, Mangelings D, Vander Heyden Y. Chiral separations in normal phase liquid chromatography: Enantioselectivity of recently commercialized polysaccharide-based selectors. Part I: Enantioselectivity under generic screening conditions. Journal of Pharmaceutical and Biomedical Analysis. 2011;**55**:414-423

[57] Chankvetadze B. Recent trends in preparation, investigation and application of polysaccharide-based chiral stationary phases for separation of enantiomers in high-performance liquid chromatography. TrAC - Trends in Analytical Chemistry. 2020;**122**:115709

[58] Wang X, Li H, Quan K, Zhao L, Li Z, Qiu H. Anhydride-linked β-cyclodextrin-bonded silica stationary phases with enhanced chiral separation ability in liquid chromatography. Journal of Chromatography. A. 2021;**1651**:462338

[59] Wu C-S, editor. Handbook of Size Exclusion Chromatography. New York: M. Dekker; 1995

[60] Bouvier ESP, Koza SM. Advances in size-exclusion separations of proteins and polymers by UHPLC. TrAC - Trends in Analytical Chemistry. 2014;**63**:85-94

[61] Urh M, Simpson D, Zhao K. Affinity chromatography: General methods, in methods in enzymology, volume 463. Elsevier. 2009;**2009**:417-438

[62] Schiel JE, Mallik R, Soman S, Joseph KS, Hage DS. Applications of silica supports in affinity chromatography. Journal of Separation Science. 2006;**29**:719-737

[63] Sproβ J, Sinz A. Monolithic media for applications in affinity chromatography. Journal of Separation Science. 2011;**34**:1-16

[64] Hermanson GT, Mallia AK, Smith PK. Immobilized Affinity Ligand Techniques. Ney York: Academic Press; 1992

[65] Chi J, Zhu D, Chen Y, Huang G, Lin X. Online specific recognition of mycotoxins using aptamer-grafted ionic affinity monolith with mixed-mode mechanism. Journal of Chromatography A. 2021;**1639**:461930

[66] McGettrick A, Worrall M. Dye-ligand affinity chromatography. In: Cutler P, editor. Methods in Molecular Biology. Vol. 244. Totowa: Humana Press; 2004. pp. 151-157. DOI: 10.1385/1-59259-655-x:151

[67] Verzele D, Lynen F, De Vrieze M, Wright AG, Hanna-Brown M, Sandra P. Development of the first sphingomyelin biomimetic stationary phase for immobilized artificial membrane (IAM) chromatography. Chemical Communications. 2012;**48**:1162-1164

[68] Carrasco-Correa EJ, Ruiz-Allica J, Rodríguez-Fernández JF, Miró M. Human artificial membranes in (bio) analytical science: Potential for in vitro prediction of intestinal absorption - a review. TrAC - Trends in Analytical Chemistry. 2021;**145**:116446

[69] Kazarian AA, Barnhart W, Long J, Sham K, Wu B, Murray JK. Purification of N-acetylgalactosamine-modified-oligonucleotides using orthogonal anion-exchange and mixed-mode chromatography approaches. Journal of Chromatography A. 2022;**1661**:462679

[70] Wang L, Wei W, Xia Z, Jie X, Xia ZZ. Recent advances in materials for stationary phases of mixed-mode high-performance liquid chromatograph. TrAC - Trends in Analytical Chemistry. 2016;**80**:495-506

[71] Zhou J, Ren X, Luo Q, Gao D, Fu Q, Zhou D, et al. Ionic liquid functionalized β-cyclodextrin and C18 mixed-mode stationary phase with achiral and chiral separation functions. Journal of Chromatography A. 2020;**1634**:461674

[72] Sykora D, Rezanka P, Zaruba K, Kral V. Recent advances in mixed-mode chromatographic stationary phases. Journal of Separation Science. 2019;**42**:89-129

[73] Fan F, Lu X, Liang X, Wang L, Guo Y. Preparation of hydrogel nanocomposite functionalized silica microspheres and its application in mixed-mode liquid chromatography. Journal of Chromatography A. 2022;**1662**:462745

[74] Yang F, Bai Q, Zhao K, Gao D, Tian L. Preparation of a novel weak cation exchange/hydrophobic interaction chromatography dual-function polymer-based stationary phase for protein separation using "thiol-ene click chemistry". Analytical and Bioanalytical Chemistry. 2015;**407**:1721-1734

[75] Zheng Y, Wan M, Zhou J, Luo Q, Gao D, Fu Q, et al. Striped covalent organic frameworks modified stationary phase for mixed mode chromatography. Journal of Chromatography A. 2021;**1649**:462186

[76] Liu D, Wang H, Liang M, Nie Y, Liu Y, Yin M, et al. Polymerized phosphonium ionic liquid functionalized silica microspheres as mixed-mode stationary phase for liquid chromatographic separation of phospholipids. Journal of Chromatography A. 2021;**1660**:462676

[77] Li H, Liu C, Zhao L, Xu D, Zhang T, Wang Q, et al. A systematic investigation of the effect of sample solvent on peak shape in nano- and microflow hydrophilic interaction liquid chromatography columns. Journal of Chromatography A. 2021;**1655**:462298

Chapter 2

Perspective Chapter: Mixed-Mode Chromatography

Ngoc-Van Thi Nguyen

Abstract

In this chapter, we present mixed-mode stationary phases and their applications in the determination of nonpolar, polar, and charged compounds, as well as larger molecules such as peptides or proteins using a single column. Mixed-mode chromatography (MMC) has been growing rapidly in recent years, owing to the new generation of mixed-mode stationary phases and a better understanding of multimode interactions. Mixed-mode chromatography provides a wide range of selectivities and adequate retention of a variety of compounds, especially polar and charged molecules. In summary, this technique is particularly useful in the pharmaceutical analysis of drugs, impurities, biopharmaceuticals, and polar compounds in natural products.

Keywords: mechanisms, intermolecular interactions, mixed-mode chromatography, mixed-mode stationary phases

1. Introduction

The development of liquid chromatography is one of the most active areas of research in separation science, with applications in various fields, such as drug analysis, medicinal chemistry, agriculture, food chemistry, and bioanalysis. This study aims to determine the optimal working conditions for the effective and selective separation of chemical compounds. In the research process, the choice of optimal chromatography conditions is of prime importance, including the determination of a suitable liquid chromatography mode and the investigation of mobile phase characteristics (pH, type of organic modifier, mobile phase additive, etc.) [1].

Chromatographic retention processes can be divided into many types, such as normal-phase, reversed-phase, ion-exchange, hydrophobic interaction, hydrophilic interaction, and metal coordination chromatography. These chromatographic methods are known as single-mode chromatography because the retention of solutes in these chromatograms is dependent on a single-retention mechanism. For instance, in reversed-phase chromatography, problems may be encountered during the analysis of highly polar (or charged) compounds. Hydrophilic interaction chromatography (HILIC) is designed for the analysis of polar compounds; however, it is still affected by a range of challenges, such as low solubility in highly organic media, the amount of organic solvents used, the sample matrix that affects retention, and the retention extent of hydrophobic analytes that can be controlled. Ion-exchange chromatography can be used for charged molecules, but not for neutral analytes. Therefore, mixed-mode chromatography (MMC) can be utilized to resolve some of the problems associated with each of the other mechanisms [2].

Mixed-mode chromatography (MMC) separates solutes by using a stationary phase that involves in the separation two or more types of interactions. Compared to single-mode chromatography, mixed-mode chromatography can simultaneously act on different functional groups of the solute, such as hydrophobic and ionic groups [3]. Mixed-mode chromatography is not a new technique. Many chromatographic matrices are based on rigid supports, such as cellulose, agarose, polyacrylamide, or silica gel, which are modified to produce specific functionalities on their surfaces. If the solute is a substance with numerous functional groups, such as amino acids, nucleic acids, peptides, and proteins, which are commonly found in biological samples, mixed-mode chromatography will exhibit a distinct behavior as opposed to that of single-mode chromatography [4].

Recently, MMC has been receiving increasing attention as an alternative or complementary tool to traditional chromatography (reversed phase, ion exchange, and normal phase) in pharmaceutical and biopharmaceutical applications because of its efficient selectivity and adequate retention of a variety of compounds—particularly polar and charged molecules. To achieve better solute retention characteristics, selectivity, and separation capabilities, mixed-mode stationary phases must be designed and synthesized based on the specific structural characteristics of different compounds. Additionally, the diversity of the mixed-mode stationary phase depends on the diversity of the analyte structure and its properties. It is expected that the applications of mixed-mode chromatography will increase in the future and serve as a power resolution for the separation and purification of biological substances.

2. Mechanisms of mixed-mode chromatography

2.1 Mechanisms related to stationary phases

2.1.1 Classification of stationary phases by chemistry design

In MMCs, stationary phases have been prepared using several types of stationary phases involving different mechanisms. According to the study design, mixed-mode stationary phases can be divided into four categories [2, 3].

Type 1: A mixed-mode stationary phase is created by combining two types of stationary phase particles (each with a single chemistry) and packing them into a single column (**Figure 1**). However, the major drawbacks of this approach are the non-homogeneity of the stationary phases and low batch-to-batch reproducibility.

Type 2: The surface of the stationary phase is modified with a mixture of ligands of different chemistries. This is a second-generation approach, but its disadvantages are similar to those of Type 1. Thus, Types 1 and 2 are not commonly used because of their performance limitations.

Types 3 and 4: Embedded (Type 3) and tipped ligands (Type 4) are the third-generation mixed-mode phases that improve the reproducibility and homogeneity of the stationary phase. The functional groups (polar or ionic groups) of the embedded ligands are close to the pore surface, and the hydrophobic parts of the ligands extend to the mobile phase. In contrast, the tipped ligands have functional groups at the ends of the hydrophobic chains.

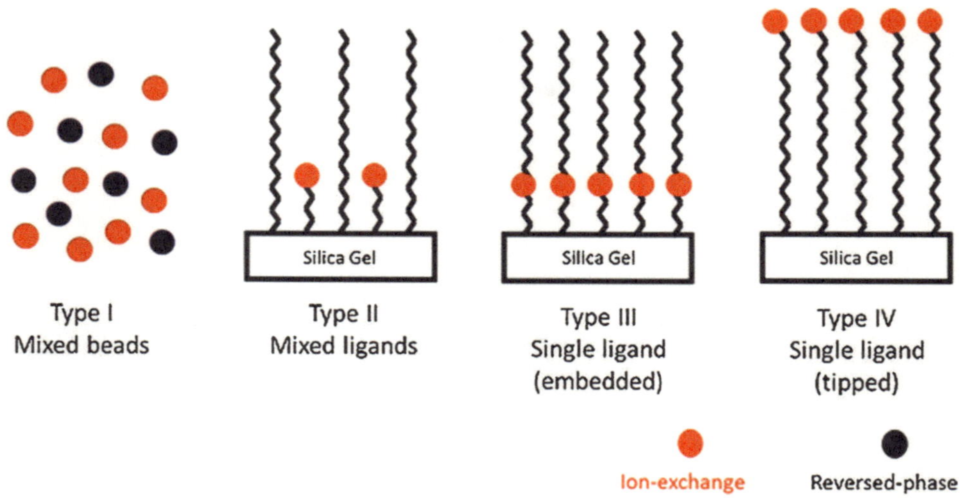

Figure 1.
Types of mixed-mode stationary phases classified by chemistry designs [5].

2.1.2 Combinations of separation modes in MMC

2.1.2.1 Reversed-phase ion-exchange stationary phases (RP-IEX)

Polar compounds, such as biologically active molecules, natural products, and drug metabolites containing several functional groups, tend to be weakly retained in the reversed phase, resulting in poor separation. With the combination of hydrophobic and ion-exchange mechanisms in the mixed-mode stationary phases, the selectivity and retention of both the hydrophobic and polar compounds are improved [4]. In addition, it is the most popular ligand in MMC and is mainly used for the separation of peptides, nucleotides, basic drugs, and their metabolites. The ligands consist of a hydrophobic part (alkyl chains or aromatic hydrocarbons) and an ionic part embedded in the end, middle, or vicinity of the hydrophobic part. Depending on the structure of the ionic part, four ion-exchange modes can be classified: quaternary amines are used as strong cation-exchange groups (SCX); primary, secondary, or tertiary amines are used as weak cation-exchange groups (WCX); sulfonic acids are used as strong anion-exchange groups (SAX); and carboxyl groups are used as weak anion-exchange groups (WAX) (**Figure 2**) [6]. The retention mechanism of this mixed-mode phase was based on the formation of a divalent complex involving hydrophobic and oppositely charged analytes. Moreover, repulsive ionic interactions with identically charged functional groups also affect analyte retention in mixed-mode stationary phases. Thus, separation can be optimized by adjusting the mobile phase parameters, such as pH, ionic strength (including the concentration of buffers and modifiers), and solvent strength [4]. For example, the C18/SAX column shows strong retention of acidic compounds due to electrostatic attraction under basic conditions. In addition, the retention values increase with increasing alkyl chain length of the analytes. Elution can be achieved using acidic conditions with high percentages of organic solvents, and/or high ionic strength to make neutral the acidic compounds and weaken the hydrophobic interactions. In contrast, the RP/SCX and RP/WCX columns can effectively retain base compounds, such as peptides

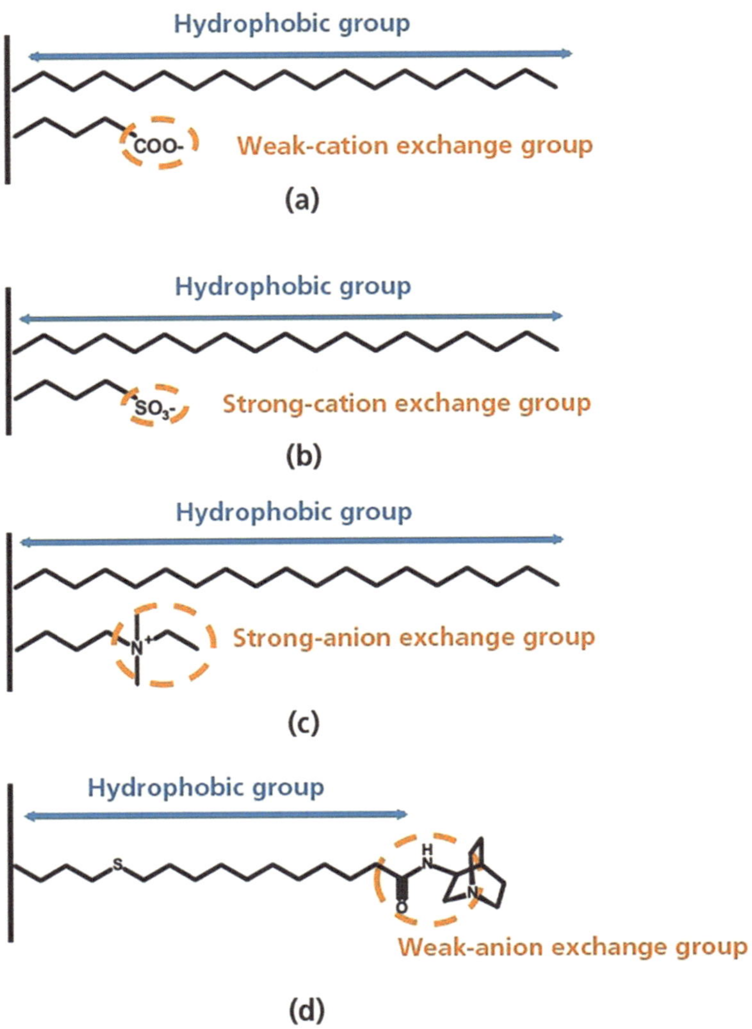

Figure 2.
Structures of some RP-IEX mixed-mode stationary phases: (a) RP-WCX, (b) RP-SCX, (c) RP-SAX, and (d) RP-WAX [6].

and alkaloids, under acidic conditions. If a mobile phase with a neutral pH and a low ionic strength is used, the retention of these compounds is strongly influenced by hydrophobic interactions [7].

In the field of mixed-mode reversed-phase/ion-exchange stationary phase, mixed-mode RP/AX based on ethylene-bridged hybrid (BEH) organic/inorganic particles was recently developed, named Atlantis BEH C18 AX. The intermediate C18 surface concentration (1.6 μmol/m^2) together with tertiary alkylamine groups) makes BEH C18 AX compatible with highly aqueous mobile phases. The BEH particles used for the BEH C18 AX stationary phase have an average pore diameter of 95 Å that increases retention, stemming from the 46% higher surface area. Furthermore, hydrophilic anion-exchange group of them create positive surface charge, which show stronger retention of negatively charged compounds in a wider pH range while using with buffers of pH 3.0–6.9 in the survey. The extended upper pH limit of BEH C18 AX allows it to be used with a wider range of mobile phase pH values. For samples containing ionizable analytes, mobile phase pH has been demonstrated to be a key variable to use in optimizing RP separations [8, 9].

2.1.2.2 Reversed-phase hydrophilic stationary phases (RP-HILIC)

RP-HILIC mixed-mode stationary phases have shown advantages in the separation of both hydrophobic and hydrophilic compounds, especially proteins. This combination is equivalent to combining the HILIC properties with the reversed-phase properties to analyze complicated compounds and matrices with a wide range of polarities in a single run. The ligands are composed of hydrophobic and polar groups. The hydrophobic parts can be alkyl or aromatic groups, and the hydrophilic parts can be charged or neutral functional groups, such as diol, amide, cyano, and ionic groups (**Figure 3**) [10]. For compounds with hydrophilic and polar parts, ligands containing nonpolar and polar groups can interact separately with their corresponding nonpolar and polar groups. Therefore, it is possible to improve analyte retention and separation selectivity through multivalent effects, including hydrophobic and hydrophilic interactions. In recent years, many mixed-mode stationary phases have been synthesized and applied to the analysis of surfactants, peptides, nucleotides, and proteins (**Table 1**).

2.1.2.3 Hydrophilic ion-exchange stationary phases (HILIC-IEX)

The combination of hydrophilic and ion-exchange groups presented strong advantages for analyzing charged polar compounds. The multivalent effects of these mixed-mode phases provide unique selectivity, higher retention efficiency, and a wider range of application than any single-mode phase for peptide analysis [16]. The main application of this combination is the separation of proteins and peptides. In this mode, the retention mechanism of polar compounds depends on the percentage of an organic solvent (such as acetonitrile (ACN)) in the mobile phase. If a mobile phase has a low percentage of the organic solvent, the analyte retention is dominated by ion-exchange mechanisms. An increase in the percentage of acetonitrile promoted more hydrophilic interactions than ionic ones. At a high concentration of acetonitrile, the electrostatic interactions decreased significantly, whereas the hydrophilic interactions dominated the analyte retention. Bo et al. prepared a HILIC-IEX phase with adjustable selectivity to separate nucleosides and β-agonists, which were synthesized by controlling the mixture ratio of the two functional monomers [17]. In addition, the use of ionic liquids to develop HILIC-IEX stationary phases can provide an environment for multiple interactions, such as electrostatic, dipole-dipole, and π-π interactions, and hydrogen bonding. Quiao et al. developed a new HILIC-SAX phase by using glucaminium-based ionic liquids to separate nucleosides [18]. According to studies reported by Mant et al. [16], the hydrophilic cation-exchange column (HILIC-CEX) has a higher separation efficiency than the RP-LC for peptide analysis, and the highly charged peptides are best resolved by this column [16]. Hartmann et al. [19] separated amphipathic α-helical peptides using a HILIC-CEX column and an RP-LC column. Both columns presented an adequate efficiency but displayed different selectivities. With the HILIC-CEX column, the temperature had a stronger influence on the separation of peptide columns than that with the RP-LC column. The results showed that both the resolution and retention of peptides in the HILIC-CEX phase significantly improved with increasing temperature [19].

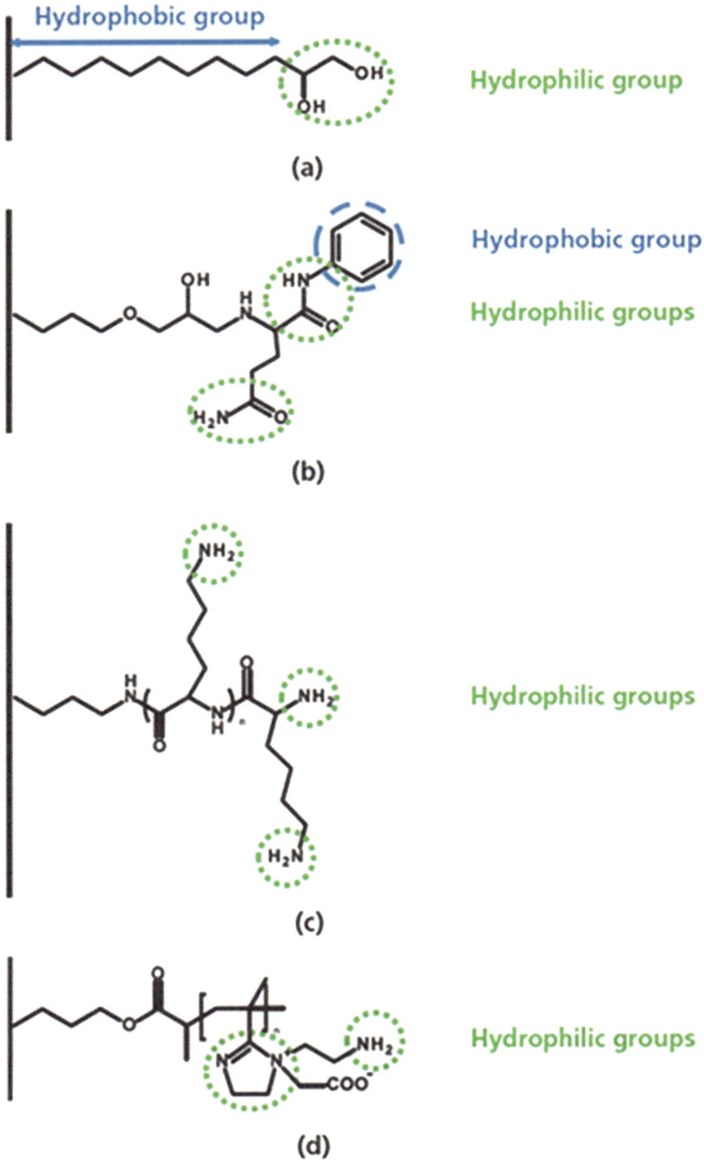

Figure 3.
Structures of some RP-HILIC mixed-mode stationary phases with (a) diol, (b) amide, and (c) amine polar groups [6].

2.1.3 Other combinations

Inclusion hydrophobic mixed-mode: The ligands are composed of hydrophobic parts and cavities, cages, or cryptates, which form an inclusion complex with the analytes. Thus, the multivalent effects of this mode include both the inclusion complexation and the hydrophobic interactions. A representative example of this combination is crown ether immobilized on a solid matrix. A tripartite hydrogen bond can form between the six oxygen atoms of the crown ether and three hydrogen atoms of the protonated primary amine. Therefore, these ligands can be used to retain and separate the primary amines or other protonated molecules. Additionally, hydrophobic

Hydrophobic part	Hydrophilic part	Application	References
Alkyl chain	Glycol terminus groups	Nonionic ethoxylated surfactants: of alkylphenol ethoxylates and fatty alcohol ethoxylates	[11]
Alkyl chain	Amide groups	Nucleosides and phenolic compounds	[12]
Alkyl chain of L-lysine	Terminal amine group of L-lysine	Aniline compounds (aniline, 2-nitroaniline, 4-aminophenol, 1-amino-2-methylbenzene, and 2,4-dinitroaniline)	[13]
Alkyl chain of small peptide	Amine and amide groups of small peptide Boc-Phe-Aib-Phe	Polycyclic aromatic hydrocarbons, steroids (hydrophobic compounds), nucleosides (hydrophilic compounds)	[14]
β-Hydroxyl fatty acid	Surfactin, a peptide loop including seven amino acids	Chiral separation	[15]

Table 1.
Applications of reversed-phase hydrophilic stationary phases.

interactions can form between the methylene on the crown ether and the alkyl chain of the analytes, resulting in improvement of analyte retention [4].

Inclusion hydrophilic mixed-mode: In this mode, the ligands are composed of a cavity, cage, or cryptate, and a polar group. An example of this chromatography process is the binding of crown ethers with primary amines. The hydrogen bonding between the primary amines and crown ether can be enhanced with a polar organic solvent mobile phase, leading to enhanced inclusion effects.

π-π hydrophilic mixed-mode: The ligands are designed by combining two groups: a π-electron donor or π-electron acceptor group, and a polar group. In π-π interactions, the electron-rich π system (π-electron donor) can interact with the electron-deficient π system or other π-electron acceptor groups, through electrostatic interactions. In this chromatography process, the π-interacting groups of ligands can interact with the π-interacting groups of analytes through π-π interactions, and the polar parts of the ligands can interact with the polar parts of the analytes through hydrogen bonding and/or dipole-dipole interactions. The main application of this combination is the separation of chiral compounds. An example of this mode is Pirkle-type chromatography ligands [20].

Π-π ion-exchange mixed-mode: In this mode, the stationary phase can interact with analytes through π-π interactions, dipole-dipole interactions, van der Waals forces, and electrostatic interactions. A representative example of this combination is the cinchona alkaloid derivative phase developed by Lämmerhofer M and Lindner W. Depending on the structure of the derivatives, this phase can exhibit many types of separation, such as anion exchange, cation exchange, and amphoteric ion exchange for chiral chromatography. Therefore, the main application of this phase is the separation of chiral acids, ionic chiral compounds with a wide range of polarities, and amphoteric compounds such as amino acids and small peptides. In particular, the amphoteric ion exchange can also be considered as an example of a multifunctional stationary phase because this ligand can present three ion-exchange modes for chiral separation under the conditions of a polar organic mobile phase. Thus, the anion-exchange mode is utilized for chiral acid separation, the cation-exchange mode for chiral amine separation, and the zwitterion mode for amphoteric compound separation [21].

Polymeric mixed-mode: Several novel polymeric MMC sorbents have been designed specifically for the separation of proteins, mainly serum albumins and immunoglobulins (IgGs). Heterocyclic compounds are unique as MMC ligands with specific aromaticity/hydrophobicity and dissociation properties compared with common aliphatic and aromatic compounds with capability to relatively selectively interact with some proteins, albumins, also antibodies and monoclonal antibodies [22].

Capillary-channeled polymer (C-CP) fiber stationary phases: The unique shape of C-CP gives them high surface area and when packed into columns, the fibers self-align, providing a monolith-like structure with parallel channels of 1–5 μm size. In relation to the size of proteins, the C-CP fibers surface is nonporous, which significantly reduce mass transfer resistance. Thus, separation can be run at high linear velocities and at low pressures without detrimental effect on the separation efficiency. All the studied C-CP stationary phases were able to separate a BSA/hemoglobin/lysozyme mixture at high mobile phase velocity and with acceptable elution characteristics [22].

2.2 Composition of mobile phases and their effects on mixed-mode chromatography

2.2.1 Polar organic solvents

The mobile phase used in mixed-mode chromatography usually involves a polar organic solvent, water, or a buffer. The following four properties of the solvent have significant effects on the retention and separation of analytes: solvent viscosity, dielectric constant, dipole moment, and surface tension. Solvent viscosity affects the chromatography process in various ways, especially when gradient conditions are used. Firstly, an increase in the viscosity of the mobile phase is the prime reason for an increase in the backpressure. Moreover, the column efficiency is influenced by the viscosity of the mobile phase. For example, a mixed solution of methanol and water has a higher viscosity than pure methanol. As a result, it reduces the diffusion coefficient of the solutes and exhibits a slow mass transfer, leading to a reduction in the column efficiency [23]. The dielectric constant (ε) and dipole moment (μ) characterize the polar nature of the solvent. A solvent with a higher ε value is usually considered a weaker eluent in reversed-phase chromatography, whereas the dipole moment is related to solvent polarity and has important effects on the interactions between the analytes and the ligands in hydrophilic chromatography. Finally, the surface tension of a solvent can affect analyte separation. A mobile phase with higher surface tension can lead to stronger analyte retention. In addition, the UV wavelength cutoff of the solvent must also be considered when a UV-Vis detector is used to measure the concentration of the analytes [24].

In RP-IEX mixed-mode chromatography, polar organic solvents (such as methanol, acetonitrile, ethanol, and tetrahydrofuran) were used as strong eluotropic components. Furthermore, organic solvents can control the retention and elution of analytes in the chromatography process, thereby providing scope to increase the solubility of analytes in the mobile phase. An increase in the organic solvent concentration causes the polarity of the mobile phase to decrease, and the hydrophobic interactions between the analytes and the ligands decrease, resulting in a decrease in retention. According to the eluotropic strength, the order of solvents is water $<$ methanol $<$ acetonitrile $<$ propanol $<$ isopropanol $<$ tetrahydrofuran [23]. The most commonly used polar organic solvents are methanol and acetonitrile (ACN). To analyze peptides and proteins, acetonitrile is preferred over methanol because the mixed solution of acetonitrile and water has a low viscosity, leading to excellent mass transfer.

A binary mixture of organic solvents and buffers is commonly used in HILIC-IEX chromatography. An increase in the organic solvent concentration can reduce the polarity of the mobile phase, leading to a strengthening of the hydrophilic interaction between the analytes and the ligands. In contrast, decreasing the organic solvent concentration can weaken the hydrophilic interactions, facilitate ionic interactions, and lead to compound elution. Thus, the organic solvent acts as a polarity modifier. One of the most commonly used solvents is acetonitrile [4]. As acetonitrile is an aprotic solvent that does not possess a hydrogen bond donor capacity, it cannot compete with the analytes for the ligands. If the mobile phase has a high level of acetonitrile, analytes can be adsorbed to the stationary phases through polar interactions, and they can be resorbed by reducing the acetonitrile content. Therefore, in the HILIC-IEX mixed-mode, acetonitrile has an important effect on the retention and separation of analytes. At high acetonitrile levels (up to 90%, v/v), the hydrophilic interactions may dominate the electrostatic interactions, and this may become the main factor affecting analyte retention.

Furthermore, to elute proteins that are strongly bound to the mixed-mode stationary phase and are significantly affected by hydrophobic interactions, reducing the polarity of the mobile phase by increasing the organic solvent content can be used as a severe elution method instead of reducing the salt concentration.

2.2.2 Buffers and pH

In mixed-mode chromatography, buffers are usually added to the mobile phase to either maintain the pH at an almost constant value or to adjust the pH value. Buffer systems can be selected depending on the required pH range. Buffer systems are classified into two categories according to their components.

Type 1: A buffer system is composed of a weak acid and its conjugate base, or a weak base and its conjugate acid, such as acid acetic/sodium acetate or ammonium chloride. For example, when acetate buffers are used in anion-exchange mixed-mode chromatography, CH_3COO^- can participate in the ion-exchange process by binding to the positively charged ligands. Therefore, a buffer system with buffer ions having the same charge as the ligands is ideal for mixed-mode chromatography involving ion-exchange mechanisms. Positively charged buffer ions are preferred when using an anion-exchange mechanism (having positively charged ligands), and negatively charged buffer ions are recommended for the cation-exchange mechanism.

Type 2: A buffer system contains an organic amine or an amphoteric compound that can be used in both the anion-exchange and the cation-exchange chromatography. The examples of such a buffer system are N-2-hydroxyethylpiperazine-N-2′-2-ethanesulfonic acid (HEPES) and N, N-dihydroxyethylglycine (BICINE).

The mobile phase pH can influence the charged properties of the analytes and the nature of the ligands; therefore, it can be used to promote the adsorption and elution of the target compounds [24]. Certain rules are applicable for selecting a suitable pH value for the mobile phase. Firstly, the ideal pH should be selected according to the pKa of the analytes and the ionic groups of the ligands. For example, a target compound with amine groups will be positively charged when the pH value is lower than its pKa, thus resulting in adsorption by cation-exchange ligands. Generally, for the adsorption process, the pH should be selected to charge the analytes and facilitate the electrostatic interactions between the analytes and the oppositely charged ligands.

Therefore, the pH should be lower than the pKa of the analytes by approximately 1–2 pH units when the adsorption is carried out on a cation-exchange ligand, while the pH should be higher than the pKa of analytes by approximately 1–2 pH units on anion-exchange ligands. In contrast, for the elution process, the pH should be adjusted to weaken or disrupt the interaction between the target compounds and ligands by charge repulsion. Secondly, the pH of the mobile phase should be within the stability range of the stationary phase. Finally, for protein analysis, it is necessary to select a pH value at which the proteins are stable and retain their biological activity [4].

In mixed-mode chromatography involving ion-exchange mechanisms, the charged properties of weakly acidic and basic ligands can be significantly affected by the pH value. For example, if the mixed-mode stationary phase contains weakly basic groups when the pH of the mobile phase is higher than the pKa of the ligand, then the ligand is neutrally charged and its hydrophobicity increases. Contrastingly, the ligand is positively charged and has a high hydrophilicity when the pH is lower than the pKa of the ligand. In the elution stage, pH gradient changes can be utilized to obtain a higher selectivity of the separation when the change in solvent polarity and the change in ionic strength produce no improvement in the separation efficiency. By changing the pH, the analytes and the ligands can have the same charge; therefore, the analytes can be eluted by charge repulsion. For example, in the experiment of Hostein et al. [25], α-Lactalbumin, β-lactoglobulin A, and trypsin inhibitor with pIs (protein's isoelectric point) of 4.5, 5.1, and 4.5, respectively, were separated by using a linear pH gradient from pH 3.8 to 8.0 (0.05 pH units/min) on the multimodal cation exchanger Capto MMC. When the pH value of the mobile phase was higher than that of the pIs, these proteins were negatively charged, as with the ligands, and eluted by electrostatic repulsion. The farther the pH value is from the pIs, the more negatively charged these proteins are, leading to their stronger hydrophilicity. It was observed that a shallower gradient (0.05 pH units/min to 0.01 pH units/min) reduces the sharpness of the peaks but improves the protein resolution.

2.2.3 Salts

In MMCs, salts are usually added to the mobile phase to adjust their ionic strength. Sodium chloride is usually used in the ion-exchange mode, whereas salts with higher solubility in organic solvents (such as sodium perchlorate and ammonium perchlorate) are preferred in the hydrophilic mode. In the hydrophobic mode, salts are classified into two categories: salting-out and salting-in. Salting-out salts, such as sodium sulfate, ammonium sulfate, and potassium sulfate, can be used to stabilize proteins and promote hydrophobic interactions between the proteins and the ligands. In contrast, salting-in salts, such as calcium chloride, magnesium chloride, and zinc nitrate, can increase the solubility of the proteins in water and promote protein denaturation and unfolding [26].

The ionic strength of the mobile phase has a significant effect on the retention and the elution of analytes in both the ion-exchange and the hydrophobic modes. In the mixed-mode chromatography involving ion-exchange mechanisms, because an increase in the ionic strength can suppress the electrostatic interaction between the analytes and the charged groups of the ligand, it may result in the weakening of analyte binding on the ligands, thereby leading to a decrease in the analyte retention or elution [24]. Moreover, in the hydrophobic mode, the increase in the ionic strength of the mobile phase can cause the analytes to strengthen their binding to the hydrophobic parts of the ligands, leading to an increase in the analyte retention. Hydrophobic

Additive	Function	Mechanism	References
Magnesium chloride	Improving recovery and maintaining biological activity of proteins	Promoting protein dissolution	[27]
Ethylene glycol	Reducing hydrophobic interactions	Causing a slight increase in electrostatic interactions	[28]
Glycerol	Stabilizing proteins	Inhibiting protein unfolding Interacting with hydrophobic surface regions of proteins	[29]
Urea	Affecting to hydrophobic and hydrogen bonding interactions Causing a slight decrease in electrostatic interactions	Interacting with the polar side chain and backbone of proteins Changing the solvation of proteins	[30]
Arginine	Reducing hydrophobic and electrostatic interactions and hydrogen bonding	Interaction with the polar side chain and aromatic moieties of proteins	[31]
Caprylic acid	Causing a large decline in the retention of proteins	Binding to the region which is also the binding site of proteins to mixed-mode ligands	[32]

Table 2.
Commonly used additives in mixed-mode chromatography process.

mixed-mode stationary phases are typically used for protein separation. In this mode, the proteins are adsorbed at high salting-out salt concentrations and eluted at low salt concentrations. Therefore, reducing the salt concentration in the mobile phase can be used in the elution mode [4].

2.2.4 Other additives

For protein separation, proteins can be eluted with ease by changing the pH or by reducing the polarity or the ionic strength of the mobile phase. However, when the proteins are firmly bound to the ligand, the recovery and the biological activity of the protein can decrease during the elution step. Therefore, additives are usually added to the mobile phase to reduce its polarity, resulting in the weakening of protein binding to the hydrophobic parts of ligands, and leading to an enhanced protein recovery. Some of the commonly used additives and their functions are listed in **Table 2**.

3. Pharmaceutical analysis application of mixed-mode stationary phases

3.1 Drugs and impurities in drugs

A mixed-mode column with a stationary phase of 50% hydrophobic C_{18} phase and 50% strong cation exchanger allows for a simultaneous detection of the ionic and hydrophobic analytes [33]. Acetaminophen and its related impurities, which ionize based on the mobile phase pH, are often separated for drug examination using ion-pair chromatography, which is a technique for organic charged compounds. Despite its numerous advantages, the corrosive effect of a large number of counter-ions on the stationary phase of the column is a practical drawback of the ion-pair chromatography. As a result, the mixed-mode stationary phases can overcome the limitations of ion-pair chromatography, allowing for the simultaneous separation of ionic and neutral organic molecules without practical constraints [34]. Furthermore, because of the lack of UV chromophores in most drugs, refractive index (RI) and evaporative light-scattering detection (ELSD) detectors have been utilized. However, these approaches are insensitive or have compatibility issues with gradient elution. Recently, charged aerosol detection (CAD) has been developed as a new type of detector for high-performance liquid chromatography (HPLC) applications. CAD is a universal detection technique for nonvolatile and semi-volatile substances with higher sensitivity and reproducibility than other types of detectors. It is highly convenient in usage as it eliminates the necessity for parameter optimization [35]. **Table 3** shows the combinations of MMC and CAD detectors, as well as the applications of various types

Drugs		
Compounds	**Mechanisms**	**References**
Atovaquone, proguanil, and its two main metabolites	50% C_{18} phase and 50% strong cation exchanger	[33]
Naproxen sodium and adenine hydrochloride An undisclosed drug in hemifumarate salt form	Acclaim Trinity P1 stationary phase with CAD detector	[35]
Imidazole, pyrazole, pyridine, pyridazine, piperidine	Reversed-phase and ion-exchange characteristics	[36]
Etidronate disodium [(1-hydroxyethylidene) bisphosphonate]	Primesep SB column (anion-exchange reverse phase column) with CAD detector	[37]
Flurbiprofen, flufenamic acid, mefenamic acid, ibuprofen, loxoprofen, ketoprofen, carprofen, indoprofen sulindac	C_{18}-DTT stationary phase (dithiothreitol silica (SiO_2) A reversed-phase liquid chromatography/hydrophilic interaction liquid chromatography	[38]
metoprolol, salicylic acid, acetylsalicylic acid, propranolol, betamethasone, imipramine, clotrimazole, thioridazine, indomethacin flurbiprofen	Two UHPLC mixed-mode hybrid CSH (charged surface hybrid) stationary phases modified by C_{18} or Phenyl group	[39]

Impurity			
Compounds	**Impurity analysis**	**Mechanisms**	**References**
Acetaminophen (paracetamol)	4-Nitrophenol, 4′-chloroacetanilide 4-Aminophenol P-Benzoquinone Hydroquinone	Octadecylsilane/strong cation exchanger (C_{18}/SCX)	[34]
L-Methionine	N-Acetyl-dl-methionine N-Acetyl-l-methionyl-l-methionine N-Acetyl-l-methionyl-d-methionine and its enantiomers L-Methionine-sulfoxide	Reversed phase/cation exchange	[40]
Bispecific IgG	Homodimer Impurities	Mixed-mode size exclusion chromatography (mmSEC)	[41]

Table 3.
Application of mixed-mode stationary phases in drugs and impurity.

of MMCs for drugs and impurities. Therefore, it is also a viable analytical tool for concurrently determining a wide range of drugs, pharmaceuticals, and their related compounds in a particular procedure [33].

3.2 Metabolomics applications

A common target of pharmacokinetic studies is the development of a biological analysis method for simultaneous observation of a wide range of drugs in a biological matrix. The tandem usage of reversed-phase and ion-exchange chromatography in MMCs has shown favorable results on the retention of polar and nonpolar small molecules in a single run [42]. In addition, efficient retention and separation of the above compounds were obtained under common and MS-friendly RP conditions, reaching a high point of selectivity and sensitivity. Therefore, MMC tandem mass spectrometry has been commonly applied in metabolic analysis. For instance, the study by Roverso et al. [43] demonstrated the effective retention of selected highly polar metabolites, which was performed by using a mixed cationic-RP column, and simultaneously obtained an efficient separation in the analysis without ion pair and derivatization of 2,4-diaminobutyric acid (DAB) and isobaric beta-methylamino-L-alanine (BMAA) [43]. The metabolomics applications are summarized in **Table 4**.

In addition, combination of MMC with molecular imprinting technology is also improving recognition selectivity for protein BSA, which proved a potential combination of other chromatography modes and molecular imprinting technology [22].

3.3 Biopharmaceuticals and polar compounds in natural products

For biopharmaceutical analysis, Capto and HEA HyperCel MMC ligands with multimodal functionality have been commercialized. Capto includes a carboxyl group that exhibits the characteristics of a phenyl group involved in hydrophobic interactions, and a weak cation exchanger. HEA HyperCel contains a hexyl group that is involved in hydrophobic interactions, and a protonable amine localized in the spacer arm. The application of this type of MMC has been demonstrated in the research by Sophie Maria et al. for mAb determination [46]. Meanwhile, tri-mixed-mode

Metabolites	Matrices	Mechanisms	References
Cytarabine (ara-C)	Mouse plasma	Reversed phase/ion exchange	[42]
Trimethylamine N-oxide (TMAO) Beta-methylamino-L-alanine (BMAA) 2,4-Diaminobutyric acid (DAB)	Plasma and urine	Reversed phase/cation exchange	[43]
S-Propargyl-cysteine (SPRC)	Rat plasma	Reversed phase/cation exchange	[44]
Phosphorylated carbohydrates	—	Reversed phase/weak anion exchange, combine with a charged aerosol detector	[45]

Table 4.
Metabolomics application of mixed-mode stationary phases.

Compounds	Mechanisms	References
Biopharmaceuticals		
Recombinant monoclonal antibodies (mAbs)	Capto MMC resin	[46]
Model drug product was made by mixing an IgG1 mAb (MW 145 kDa, pI 8.4) with seven excipients from different property categories: sodium and potassium (cation), chloride and succinate (anion), histidine (zwitterions), trehalose (hydrophilic neutral sugar), and PS80 (hydrophobic nonionic surfactant).	Combination of four different separation mechanisms (size exclusion, anion exchange, cation exchange, and reverse phase)	[47]
Pembrolizumab (Anti-PD1 IgG4 wild type and S228P mutant)	Weak hydrophobic interactions and size exclusion	[48]
Polar compounds in natural products		
Theophylline, gastrodin, tetrahydropalmatine, lycorine, berberine, sinomenine, and tetrandrine	C18-DTT stationary phase [A reversed-phase liquid chromatography (RPLC)/hydrophilic interaction liquid chromatography (HILIC)]	[42]
H. diffusa and S. barbata aqueous extract	Strong anion exchange and reversed phase	[49]
Amino acids (L-pyroglutamic Acid, L-valine, L-tyrosine, L-proline) Carbohydrates (D-glucose, D-sucrose, α-D-glucopyranosyl-(1 → 2)-βD-fructofuranosyl-(1 → 1)-α-D-galactopyranose, and D-stachyose) Succinic acid (Dan-Qi pair that make from Radix Salvia miltiorrhiza and Radix Panax notoginseng)	Directly coupled reversed-phase and hydrophilic interaction liquid chromatography–tandem mass spectrometry	[50]

Table 5.
Application of mixed-mode stationary phases in biopharmaceuticals and polar compounds in natural products.

chromatography and another dual combination of MMC are also useful tools for biopharmaceutical analysis. The applications are listed in **Table 5**.

Table 5 also illustrates the determination of polar compounds in natural products. Strong anion-exchange and reversed-phase mechanisms were analyzed in both the polar and more apolar ionic and nonionic compounds and have been used to determine Chinese herbal medicines that provide good retention for separation [49]. Thus, the combination of reversed phase and hydrophilic interactions is a common mechanism in this field because of its suitable characteristics for the detection of polar compounds, especially those of natural origin.

4. Conclusion

In this chapter, advanced applications of mixed-mode stationary phases are reviewed. By adjusting the ratio of organic matter and the mobile phase concentration, the reversed-phase, HILIC, and IEX modes can be successively used. In conclusion, RP-IEX, RP-HILIC, and HILIC-IEX are the most commonly preferred mixed-mode stationary phases.

Acknowledgements

We would like to express their hearty gratitude to Can Tho University of Medicine and Pharmacy. We also thank all of our colleagues for their excellent assistance.

We would like to thank Editage (www.editage.com) for English language editing.

Author details

Ngoc-Van Thi Nguyen
Can Tho University of Medicine and Pharmacy, Can Tho City, Vietnam

*Address all correspondence to: nguyenthingocvanct@gmail.com; ntnvan@ctump.edu.vn

References

[1] Taraji M, Haddad PR, Amos RI, Talebi M, Szucs R, Dolan JW, et al. Chemometric-assisted method development in hydrophilic interaction liquid chromatography: A review. Analytica Chimica Acta. 2018;**1000**:20-40

[2] Taylor T. Mixed-mode HPLC separations: What, why, and how. LCGC North America. 2014;**32**:226

[3] Zhang K, Liu X. Mixed-mode chromatography in pharmaceutical and biopharmaceutical applications. Journal of Pharmaceutical and Biomedical Analysis. 2016;**128**:73-88. DOI: 10.1016/j.jpba.2016.05.007

[4] Wan QH. Mixed-mode Chromatography Principles, Methods, and Applications. Chapter 3. Berlin, Germany: Springer; 2021. pp. 71-117

[5] Liu X, Pohl CA. HILIC behavior of a reversed-phase/cationexchange/anion-exchange trimode column. Journal of Separation Science. 2010;**33**:779-786. DOI: 10.1002/jssc.200900645

[6] Eric L, Caroline W, Elise L, Philippe H, Sophie B. Mixed-mode chromatography—A review. LCGC Supplements. 2017;**30**(6):22-33

[7] Lemasson E, Richer Y, Bertin S, et al. Characterization of retention mechanisms in mixed-mode HPLC with a bimodal reversed-phase/cation-exchange stationary phase. Chromatographia. 2018;**81**:87-399. DOI: 10.1007/s10337-018-3477-5

[8] Walter TH, Alden BA, Field JA, Lawrence NL, Osterman DL, Patel AV, et al. Characterization of a highly stable mixed-mode reversed-phase/weak anion-exchange stationary phase based on hybrid organic/inorganic particles. Journal of Separation Science. 2021;**44**:1005-1014. DOI: 10.1002/jssc.202001136

[9] Kadlecová Z, Kozlík P, Tesařová E, Gilar M, Kalíková K. Characterization and comparison of mixed-mode and reversed-phase columns; interaction abilities and applicability for peptide separation. Journal of Chromatography A. 2021;**1648**:462182. DOI: 10.1016/j.chroma.2021.462182

[10] Yang Y, Geng X. Mixed-mode chromatography and its applications to biopolymers. Journal of Chromatography A. 2011;**1218**:8813-8825. DOI: 10.1016/j.chroma.2011.10.009

[11] Liu X, Pohl C. New hydrophilic interaction/reversed-phase mixed-mode stationary phase and its application for analysis of nonionic ethoxylated surfactants. Journal of Chromatography A. 2008;**1191**:83-89. DOI: 10.1016/j.chroma.2007.12.012

[12] Aral T, Aral H, Ziyadanoğulları B, Ziyadanoğulları R. Synthesis of a mixed-model stationary phase derived from glutamine for HPLC separation of structurally different biologically active compounds: HILIC and reversed-phase applications. Talanta. 2015;**131**:64-73. DOI: 10.1016/j.talanta.2014.07.060

[13] Li Y, Xu Z, Feng Y, Liu X, Chen T, Zhang H. Preparation and evaluation of poly-L-lysine stationary phase for hydrophilic interaction/reversed-phase mixed-mode chromatography. Chromatographia. 2011;**74**:523-530. DOI: 10.1007/s10337-011-2120-5

[14] Ray S, Takafuji M, Ihara H. Chromatographic evaluation of a newly

designed peptide-silica stationary phase in reverse phase liquid chromatography and hydrophilic interaction liquid chromatography: Mixed mode behavior. Journal of Chromatography A. 2012;**1266**:43-52. DOI: 10.1016/j.chroma.2012.10.004

[15] Ohyama K, Inoue Y, Kishikawa N, Kuroda N. Preparation and characterization of surfactin-modified silica stationary phase for reversed-phase and hydrophilic interaction liquid chromatography. Journal of Chromatography A. 2014;**1371**:257-260. DOI: 10.1016/j.chroma.2014.10.073

[16] Mant CT, Hodges RS. Mixed-mode hydrophilic interaction/cationexchange chromatography: Separation of complex mixtures of peptides of varying charge and hydrophobicity. Journal of Separation Science. 2008;**31**:1573-1584. DOI: 10.1002/jssc.200700619

[17] Bo C, Wang X, Wang C, Wei Y. Preparation of hydrophilic interaction/ion-exchange mixed-mode chromatographic stationary phase with adjustable selectivity by controlling different ratios of the co-monomers. Journal of Chromatography A. 2017;**1487**:201-210. DOI: 10.1016/j.chroma.2017.01.061

[18] Qiao L, Wang S, Li H, Shan Y, Dou A, Shi X, et al. A novel surface-confined glucaminium-based ionic liquid stationary phase for hydrophilic interaction/anion-exchange mixed-mode chromatography. Journal of Chromatography A. 2014;**1360**:240-247. DOI: 10.1016/j.chroma.2014.07.096

[19] Hartmann E, Chen Y, Mant CT, Jungbauer A, Hodges RS. Comparison of reversed-phase liquid chromatography and hydrophilic interaction/cation-exchange chromatography for the separation of amphipathic a-helical peptides with L- and D-amino acid substitutions in the hydrophilic face. Journal of Chromatography A. 2003;**1009**:61-71. DOI: 10.1016/S0021-9673(03)00620-4

[20] Pirkle WH, Pochapsky TC. Considerations of chiral recognition relevant to the liquid chromatographic separation of enantiomers. Chemical Reviews. 1989;**89**:347-362. DOI: 10.1021/cr00092a006

[21] Lämmerhofer M, Lindner W. Quinine and quinidine derivatives as chiral selectors I. Brush type chiral stationary phases for high-performance liquid chromatography based on cinchonan carbamates and their application as chiral anion exchangers. Journal of Chromatography A. 1996;**741**:33-48. DOI: 10.1016/0021-9673(96)00137-9

[22] Sýkora D, Řezanka P, Záruba K, Král V. Recent advances in mixed-mode chromatographic stationary phases. Journal of Separation Science. 2019;**42**(1):89-129. DOI: 10.1002/jssc.201801048

[23] Snyder LR, Kirkland JJ, Glajch JL. Practical HPLC Method Development. 2nd ed. Hoboken: John Wiley & Sons, Inc; 1997

[24] Lee TD. Introduction to modern liquid chromatography. Journal of the American Society for Mass Spectrom. 2011;**22**:196. DOI: 10.1002/9780470508183

[25] Holstein MA, Nikfetrat AA, Gage M, Hirsh AG, Cramer SM. Improving selectivity in multimodal chromatography using controlled pH gradient elution. Journal of Chromatography A. 2012;**1233**:152-155. DOI: 10.1016/j.chroma.2012.01.074

[26] Kang B, Tang H, Zhao Z, Song S. Hofmeister series: Insights of ion

specificity from amphiphilic assembly and Interface property. ACS Omega. 2020;**5**(12):6229-6239. DOI: 10.1021/acsomega.0c00237

[27] Durkee KH, Roh BH, Doellgast GJ. Immunoaffinity chromatographic purification of Russell's viper venom factor X activator using elution in high concentrations of magnesium chloride. Protein Expression and Purification. 1993;**4**(5):405-411. DOI: 10.1006/prep.1993.1053

[28] Hou Y, Cramer SM. Evaluation of selectivity in multimodal anion exchange systems: A priori prediction of protein retention and examination of mobile phase modifier effects. Journal of Chromatography A. 2011;**1218**(43):7813-7820. DOI: 10.1016/j.chroma.2011.08.080

[29] Vagenende V, Yap MG, Trout BL. Mechanisms of protein stabilization and prevention of protein aggregation by glycerol. Biochemistry. 2009;**48**(46):11084-11096. DOI: 10.1021/bi900649t

[30] Zheng W, Borgia A, Buholzer K, Grishaev A, Schuler B, Best RB. Probing the action of chemical denaturant on an intrinsically disordered protein by simulation and experiment. Journal of the American Chemical Society. 2016;**138**(36):11702-11713. DOI: 10.1021/jacs.6b05443

[31] Hirano A, Shiraki K, Kameda T. Effects of arginine on multimodal chromatography: Experiments and simulations. Current Protein and Peptide Science. 2019;**20**(1):40-48

[32] Holstein MA, Parimal S, McCallum SA, Cramer SM. Mobile phase modifier effects in multimodal cation exchange chromatography. Biotechnology and Bioengineering. 2012;**109**(1):176-186. DOI: 10.1002/bit.23318

[33] Bergqvist Y, Hopstadius C. Simultaneous separation of atovaquone, proguanil and its metabolites on a mixed mode high-performance liquid chromatographic column. Journal of Chromatography B: Biomedical Sciences and Applications. 2000;**741**(2):189-193. DOI: 10.1016/s0378-4347(00)00082-7

[34] Calinescu O, Badea IA, Vladescu L, Meltzer V, Pincu E. HPLC separation of acetaminophen and its impurities using a mixed-mode reversed-phase/cation exchange stationary phase. Journal of Chromatographic Science. 2012;**50**(4):335-342. DOI: 10.1093/chromsci/bmr043

[35] Zhang K, Dai L, Chetwyn NP. Simultaneous determination of positive and negative pharmaceutical counterions using mixed-mode chromatography coupled with charged aerosol detector. Journal of Chromatography A. 2010;**1217**(37):5776-5784. DOI: 10.1016/j.chroma.2010.07.035

[36] Abdighahroudi MS, Lutze HV, Schmidt TC. Development of an LC-MS method for determination of nitrogen-containing heterocycles using mixed-mode liquid chromatography. Analytical and Bioanalytical Chemistry. 2020;**412**:4921-4930

[37] Liu XK, Fang JB, Cauchon N, Zhou P. Direct stability-indicating method development and validation for analysis of etidronate disodium using a mixed-mode column and charged aerosol detector. Journal of Pharmaceutical and Biomedical Analysis. 2008;**46**(4):639-644. DOI: 10.1007/s00216-020-02665-x

[38] Wang Q, Long Y, Yao L, Xu L, Shi ZG, Xu L. Preparation, characterization and application of a reversed phase liquid chromatography/hydrophilic interaction chromatography mixed-mode C18-DTT stationary

phase. Talanta. 2016;**146**:442-451. DOI: 10.1016/j.talanta.2015.09.009

[39] Nováková L, Vlčková H, Petr S. Evaluation of new mixed-mode UHPLC stationary phases and the importance of stationary phase choice when using low ionic-strength mobile phase additives. Talanta. 2012;**93**:0-105. DOI: 10.1016/j.talanta.2012.01.054

[40] Kühnreich R, Holzgrabe U. Impurity profiling of L-methionine by HPLC on a mixed mode column. Journal of Pharmaceutical and Biomedical Analysis. 2016;**122**:118-125. DOI: 10.1016/j.jpba.2016.01.057

[41] Yan Y, Xing T, Wang S, Daly TJ, Li N. Coupling mixed-mode size exclusion chromatography with native mass spectrometry for sensitive detection and quantitation of homodimer impurities in bispecific IgG. Analytical Chemistry. 2019;**91**(17):11417-11424. DOI: 10.1021/acs.analchem.9b02793

[42] Hsieh Y, Duncan CJG, Liu M. A mixed-mode liquid chromatography-tandem mass spectrometric method for the determination of cytarabine in mouse plasma. Journal of Chromatography B: Analytical Technologies in the Biomedical and Life Sciences. 2007;**854**(1-2):8-12. DOI: 10.1016/j.jchromb.2007.03.034

[43] Roverso M, Di Gangi IM, Favaro G, Pastore P, Bogialli S. Use of a mixed cationic-reverse phase column for analyzing small highly polar metabolic markers in biological fluids for multiclass LC-HRMS method. Applied Sciences. 2020;**10**(20):7137. DOI: 10.3390/app10207137

[44] Zheng Y, Liu H, Ma G, Yang P, Zhang L, Gu Y, et al. Determination of S-propargyl-cysteine in rat plasma by mixed-mode reversed-phase and cation-exchange HPLC–MS/MS method and its application to pharmacokinetic studies. Journal of Pharmaceutical and Biomedical Analysis. 2011;**54**(5):1187-1191. DOI: 10.1016/j.jpba.2010.11.027

[45] Hinterwirth H, Lämmerhofer M, Preinerstorfer B, Gargano A, Reischl R, Bicker W, et al. Selectivity issues in targeted metabolomics: Separation of phosphorylated carbohydrate isomers by mixed-mode hydrophilic interaction/weak anion exchange chromatography. Journal of Separation Science. 2010; **33**(21):3273-3282. DOI: 10.1002/jssc.201000412

[46] Maria S, Joucla G, Garbay B, Dieryck W, Lomenech A-M, Santarelli X, et al. Purification process of recombinant monoclonal antibodies with mixed mode chromatography. Journal of Chromatography A. 2015;**1393**:57-64. DOI: 10.1016/j.chroma.2015.03.018

[47] He Y, Friese OV, Schlittler MR, Wang Q, Yang X, Bass LA, et al. On-line coupling of size exclusion chromatography with mixed-mode liquid chromatography for comprehensive profiling of biopharmaceutical drug product. Journal of Chromatography A. 2012;**1262**:122-129. DOI: 10.1016/j.chroma.2012.09.012

[48] Yang X, Zhang Y, Wang F, Wang LJ, Richardson D, Shameem M, et al. Analysis and purification of IgG4 bispecific antibodies by a mixed-mode chromatography. Analytical Biochemistry. 2015;**484**:173-179. DOI: 10.1016/j.ab.2015.06.014

[49] Li D, Dück R, Schmitz OJ. The advantage of mixed-mode separation in the first dimension of comprehensive two-dimensional liquid-chromatography. Journal of Chromatography A. 2014;**1358**:128-135. DOI: 10.1016/j.chroma.2014.06.086

[50] Sun WY, Lu QW, Gao H, et al. Simultaneous determination of hydrophilic and lipophilic constituents in herbal medicines using directly-coupled reversed-phase and hydrophilic interaction liquid chromatography-tandem mass spectrometry. Scientific Reports. 2017;7:7061. DOI: 10.1038/s41598-017-07087-x

Chapter 3

Perspective Chapter: Multi-Dimensional Liquid Chromatography - Principles and Applications

Esayas Tesfaye, Tadele Eticha, Ariaya Hymete and Ayenew Ashenef

Abstract

Many complex mixtures usually constitute hundreds or even thousands of individual components of interest. Such mixtures are much too complicated to be separated for analytical duties in a reasonable period of time using only a single-dimensional chromatographic method. However, if a complex mixture is separated by an initial dimension using multi-dimensional liquid chromatography, a simpler portion of that separation is collected and goes to the second dimension. Each of these fractions will be analyzed separately, allowing exceedingly complex mixtures to be resolved in a short period of time. This chapter explains the fundamental principles, theoretical discussions as well as various applications with typical examples of multi-dimensional liquid chromatography in different fields.

Keywords: multi-dimensional liquid chromatography, reversed phase liquid chromatography, stationary phase, mobile phase, chiral separation

1. Introduction

Many complex mixtures consist of hundreds or even thousands of distinct components of interest. Some of these mixtures are so convoluted that they have never been isolated fully and may never be. Obviously, such mixes can be separated unidimensionally to some degree, but there is little possibility that all of the mixture's components will elute. They will not be separated completely even by exhausting all options of using a highly efficient column under ideal circumstances and altering the chromatographic conditions (such as solvent system composition, column temperature, and mobile phase pH) throughout the elution [1, 2].

The peak capacity of one-dimensional liquid chromatography for the analysis of complicated samples is limited. To resolve as many compounds as feasible, techniques with larger peak capacities are required. The employment of multi-dimensional chromatography could be a feasible solution to this challenge. So, two-dimensional liquid chromatography (2D-LC), which has a long history in a variety of analytical domains such as proteomic and genomic research, could be a useful technique for the thorough study of complex samples [2–5].

In 2D-LC configurations, a variety of chromatographic methods have been used, with RP being the most popular due to its greater compatibility with electrospray ionization (ESI) MS, high-resolving power, and sample desalting capability options when the first dimension demands salt gradients. Because of the good orthogonality of these two separations, the vast majority of 2D-LC analyses implemented today use Strong Cation Exchange (SCX) coupled to Reverse Phase (RP) in both on-line and off-line modes. Other 2D-LC methods, including as size exclusion chromatography (SEC), affinity purification chromatography (AFC), various types or combinations of ion exchangers, anion and cation mixed-bed exchange, and hydrophilic interaction liquid chromatography (HILIC), have emerged in recent years as promising alternatives to this combination [5–8].

Giddings in 1984 was the first to establish the theory of multi-dimensional chromatographic separations. The history of "multi-dimensional" liquid separations is almost as long as that of chromatography. The word refers to the method in which a sample is subjected to many separation mechanisms such as mobile phase modifier concentration, mobile phase p^H, and column temperature. This is designed considering the physicochemical properties of the sample components and each of which again counted as an independent separation dimension in one step. The resulting 2D system has a higher-resolving power than each single dimension when two separation systems based on different (non-correlative) retention mechanisms are coupled. One-dimensional liquid chromatography is a single-step process using only one column, while multi-dimensional liquid chromatography uses two and more than two steps and columns to separate samples. At the same time, the peak capacity, separation efficiency, sample resolution, and complexity of the method increase when the user opts from one to multi-dimensional liquid chromatography. The most popular version of multi-dimensional chromatography is the two-dimensional liquid chromatography. Early multi-dimensional separations were performed in both planar and columnar modes using only a combination of paper chromatography, electrophoresis, and gels. Chromatographic advancements have boosted separation power in terms of the number of analytes separated, but this has switched focus to the separation of highly complex mixtures such as proteomics and metabolites [3, 4].

In light of its enormous potential, a number of researchers had begun to pioneer the next major step of 2D technology isoelectric-focusing X-gel electrophoresis was evolved by O'Farrell and others into a 2D technology capable of separating over 1000 proteins. Another scientist Guiochon embarked on a project to convert 1D column LC into a 2D column method [9].

2. Principles of multi-dimensional liquid chromatography

To achieve effective separations in a comprehensive multi-dimensional LC technique, the development and optimization of it necessitates the adjustment of several parameters.

2.1 Column selectivity

When building an MDLC separation, column selectivity critically affects MD system, and orthogonality (independent separation mechanisms) and finally, peak capacity. To get the greatest possible increase in peak capacity, the columns employed in the two dimensions must have varying degrees of selectivity. When optimizing a process, a number of additional variables must be taken into consideration, including mobile phase composition, flow rate, and temperature. Because two-dimensional systems with totally non-correlated selectivities are uncommon in reality, matching and optimizing the operating circumstances in both dimensions are necessary to gain a considerable boost in resolving power [10].

2.2 Orthogonality

When MDLC separation is orthogonal, it means the two separation mechanisms are independent of each other, and provide complementary selectivities. To achieve orthogonal separation, columns employed in MDLC must be different in terms of dimensions and the composition of the stationary phase taking into account the physicochemical properties of the sample components including size and charge, hydrophobicity, and polarity. In 2DLC, the most critical and difficult decision is choosing which columns should be used as the first and second step. This decision affects the system's separation capabilities. The optimum result is attained when columns retain substances in a distinct way, resulting in a unique separation process. The larger the variation becomes in column chemistry, the higher the effectiveness of the separation process [4, 11, 12].

2.3 Peak capacity

The peak capacity of multi-dimensional separation system is the maximum number of peaks to be separated on a given column. In multi-dimensional chromatography, the peak capacities are multiplicative, which is the best assessment of performance under gradient settings. According to the product rule, in ideal circumstances a 2DLC system's overall peak capacity is the product of the peak capacities of the two dimensions [13, 14].

2.4 Resolution and sampling rate

Keeping the first-dimensional resolution is a vital criterion to follow in a complete 2D separation, which may be done by conducting a sufficient number of peak samples. To obtain the highest two-dimensional resolution, each separated peak in the first dimension should be sampled at least three times into the second dimension. The sampling time and rate affects the analysis time and its resolution. The shortest sampling time or rate into the second dimension gives the best resolution and longer sampling times decrease resolution. At the same time, the analysis time of the second-dimensional separation system is a major factor in determining the total analysis time of comprehensive two-dimensional separation systems and its resolution [15–17].

The most critical factors affecting the results of an on-line 2DLC separation are the effects of the stationary phase, mobile phase, and temperature on separation selectivity and peak capacity, compatibility of mobile phase in each dimension, and the matching of column dimensions and flow rates in each dimension. In general, excellent orthogonality across the different dimensions, great peak capacity in each dimension, preserving the early dimensions' peak capacity, and reducing sample loss throughout the process are all considered to be fundamental principles for a productive multi-dimensional design [4, 15].

MDLC Modes	First dimension	Second dimension	Characteristics
SCX-RPLC	Charge	Hydrophobicity	Orthogonality, high capacity, and efficiency, most wide application
HILIC-RPLC	Hydrophilicity/polarity	Hydrophobicity	Orthogonality, good selectivity of PTMs (post-translational modification in proteomics)
SEC-RPLC	Molecular size	Hydrophobicity	Orthogonality, low-resolving power, and high-loading capacity of SEC for sample fractionation
CIEF-RPLC	Isoelectric point	Hydrophobicity	Orthogonality, strong sample concentration of CIEF
RPLC-RPLC	Hydrophobicity under different pH conditions	Hydrophobicity under different pH conditions	Limited orthogonality, excellent resolving power

CIEF: Capillary Isoelectric focusing; SEC: Size exclusion chromatography.

Table 1.
Multi-dimensional liquid chromatography combination modes.

2.5 Combinations of separation modes in MDLC

Multi-dimensional liquid-based separation technologies have been constantly improved and innovated to get better results throughout time. MDLC may be used in a variety of ways to maximize its separation power, depending on the analytical application. Among the potential separation mode combinations are ion exchange chromatography/reversed-phase chromatography (IEC/RPC), size exclusion chromatography/reversed-phase chromatography (SEC/RPC), size exclusion chromatography/ion exchange chromatography (SEC/IEC), normal-bonded phase chromatography/reversed-phase chromatography (NPC/RPC), liquid-solid chromatography/reversed-phase chromatography (LSC/RPC), affinity chromatography/reversed phase chromatography (AC/RPC), and achiral column/chiral column. In today's 2D-HPLC/MS coupling, the most common separation strategy is strong cation exchange (SCX) in the first dimension, followed by RPLC in the second dimension. This is because SCX is the best terms of sample capacity, whereas RPLC is the most compatible column with MS. The various techniques used in multi-dimensional chromatography are described in **Table 1** [16–18].

3. Construction modes of MDLC

MDLC is often constructed using comprehensive 2DLC and heart-cutting 2DLC. A comprehensive or heart cut onto a different chromatographic column with a greater suited selectivity may redirect co-eluting components into independent eluting components. A divert valve facilitates flows and sampling elute and diverts all or portion of the analyte of interest plus co-eluted compounds from the initial column selectively to the second column for further separation and resolution. The heart-cutting MDLC is utilized to improve component separation. This is done to pre-separate targeted constituents from interfering matrices, and only specified single fractions are sent to the second dimension. It uses typical LC conditions, specifically flow rate, which is ideal for low-abundance component identification and purity analysis [14, 17, 19].

In multiple heart cuttings, more than one area of the 1D effluent is injected onto the 2D column. 2DLC is a good technique to tackle pharmaceutical issues. In pharmaceutical analysis, only certain peaks or sections of the first dimension are of relevance; hence, its eluent does not need to be transferred to the second dimension. On-line multi-heart cutting provides more versatility. In comprehensive MDLC, it is feasible to collect as much information as possible by doing a non-targeted and thorough examination of complicated samples. It utilizes 2DLC, which transfers all fractions from first into second dimension for subsequent analysis. As explained in **Table 2** below both heart cutting and comprehensive implementation of multi-dimensional liquid chromatography have its own advantages and disadvantages. One has its own superior field of application over the other [20–22].

Multi-dimensional systems may be coupled in three ways: on-line, stop-and-go, and off-line. Multi-dimensional on-line separation allows for direct transfer of fractions from one dimension to the next for further separation. On-line coupling involves coupling the second dimension to the first dimension in real time. Under this separation system, the second analysis of a single fraction should be performed within the time it takes to collect, transport, and analyze the fraction. The key benefits are the lower sample size required, less sample loss, and faster analytical times. It has more strict conditions, such as the first dimension's solvent must be a weak eluent in the second dimension, and the second dimension must be quick enough to maintain the first dimension's resolution. Using on-line setups reduces sample handling since sample movement across dimensions is continuous through switching valves, extra pumps, and trapping columns. Notably, only a few applications employ off-line 2D-LC, indicating that on-line 2D-LC is more suited and hence more desirable for pharmaceutical analysis. This is possible because, despite its benefits, off-line 2D-LC is tedious and cannot be automated [17, 18, 20, 22].

With stop and-go method, elution from the first-dimension column is prevented while a fraction is transferred to and processed on the second-dimension column, and then continued in the first-dimension. This reduces the second-dimension time limitations but increases peak parking periods, reducing first-dimensional separation

Implementation	Number of target compounds	Advantages	Disadvantages	Typical application
Heart cut (LC-LC)	+	Simple; powerful for highly targeted work	Limited to a few target samples	Targeted analysis in complex matrix (e.g., drug metabolite in serum)
Multiple heart cut (mLC–LC)	++	Amenable to more target compounds	More complex instrumentation	Quantitation of moderately complex mixture (e.g., mixture of achiral/chiral molecules)
Comprehensive (LC × LC)	+++	Most efficient way to obtain full of sample composition can see several peaks in reasonable time	Slow speed of 2D separation, long analysis time	Sample profiling/ fingerprinting (e.g., metabolomics), discovery (untargeted) type analysis

+, ++, +++ *denotes an increase in the number and complexity of target compounds of interest for the analytical work.*

Table 2.
Comparison of key attributes of the major implementations of MDLC.

efficiency. This method has been applied effectively, most importantly in multi-dimensional protein identification technology (MPIT). The benefit of a stop-and-go strategy is that the second dimension may be much longer than an online approach [19, 22–24].

In off-line approach, eluting portions are collected at regular intervals for further separation on the second dimension. Since there is no direct connection, samples may be desalted and/or recrystallized after the initial separation, making it possible to combine chromatography that is not directly compatible. There are several ways to alter the portions (dilution or concentration or dissolution in various solvents), chemically modify them, and analyze them again in order to improve their peak capacity as the second dimension's analysis duration is unrestricted. As an additional benefit, a 2DLC separation may be performed using just one-liquid chromatography. However, this method is time-consuming and requires labor-intensive sample manipulation steps, making it is more susceptible to sample loss and contamination compared with other approaches [9, 25, 26].

4. Fields of application of multi-dimensional liquid chromatography

Several multi-dimensional chromatography systems have been introduced in the last few years to improve the separation and perform an in-depth analysis of proteome and lipidomics, environmental chemicals, polymer and oligomer separation, metabolomics, and closely related medications (chiral drugs). The range of applications for MDLC is far too broad to be covered here. So, this chapter mainly focuses on pharmaceutical applications. It explains specific field of application with few practical examples. It also presents systematically gathered scientific information from a plethora of articles scattered over a wide a range of sources.

4.1 Pharmaceuticals

In the pharmaceutical industry, detection of all synthesis-related impurities and degradation products present with the active pharmaceutical ingredient are of extreme importance. High-performance liquid chromatography has been the technique of choice for many years to assess the chemical purity of drug substances and products that are widely used in the pharmaceutical industry, from research and development stages to quality control laboratories. The peak capacity and selectivity of this conventional liquid chromatography may not be sufficient to separate all substances. The implementation of MDLC is therefore highly beneficial in order to address co-elution issues and for the verification of chiral purity of the API. To further improve the probability of success, the chromatographic peaks can be analyzed on more than one orthogonal LC system. This can be performed by utilizing a second-dimensional screening module that comprises various column types, organic modifiers, and pH adjustments [11, 12, 22, 27].

4.1.1 Trace analysis

Peak co-elution is a significant problem in pharmaceutical analysis since impurities can co-elute with API or other components. One of the most applications of 2D-LC is directed toward the separation of peaks that co-elutes in conventional 1D-LC methods. This is of prime importance for peak purity assessment. This problem can be overcome using LC–LC with a heart cut of the fraction containing API and its co-eluted impurities [10, 13, 14].

4.1.2 Chiral analysis

Since many drugs are chiral, separation of them is gaining importance especially for those with two or more chiral center race mates. For separating enantiomers, heart-cutting (or multi heart cutting) liquid chromatography can be useful. The majority of research on direct chiral separations has concentrated on analytes in a basic sample matrix, and there has been relatively little research on the direct separation of medication enantiomers in biological materials using multi-dimensional chromatography. However, some publications are available on over a wide a range of sources and is presented in **Table 3** [15, 16, 47].

4.1.3 Separations of biopharmaceuticals

Biopharmaceuticals are therapeutic proteins produced *in vivo* through recombinant DNA technology and are generally used for the treatment of severe diseases, such as cancer, autoimmune disorders, and cardiovascular diseases. Several kinds of therapeutics fall within the category of protein biopharmaceuticals (hormones, growth factors, blood factors, vaccines, anticoagulants, cytokines, and others), but monoclonal antibodies represent the largest percentage of these drugs (mAbs) followed closely by mAb-related products, such as antibody-drug conjugates (ADCs). This biopharmaceutical examines latest research on using biologics to develop new drugs, vaccines, and gene therapies in the quest to realize the promise of personalized medicine [28, 48, 49].

Several researches have arisen in recent years, in response to this useful therapeutic area, using heart cutting, multiple heart cuttings, and comprehensive 2D-LC. A tryptic digest of trastuzumab was analyzed by three different 2D-LC combinations, including CEX × RPLC, RPLC × RPLC, and HILIC × RPLC, with both UV (DAD) and MS detections. The orthogonal information obtained by the application of the different LC × LC approaches allowed for assessing both the identity and purity of the sample. Similarly, the therapeutic monoclonal antibody, herceptin is characterized by different chromatographic approaches (RPLC, HIC, SEC, CEX, and HILIC). Similarly, HIC × RPLC–HRMS was performed to obtain and profile the drug-to-antibody ratio (DAR) of brentuximab vedotin in the first dimension (HIC) with an inline desalting step performed in the second dimension (RPLC) prior to the coupling with MS that allowed accurate identification of positional isomers. Another method was developed for streamlined characterization of an antibody-drug conjugate by 2D and 4D-LC/ MS. A 4D-LC/MS method (SEC-reduction-digestion-RPHPLC) was also developed to determine the levels of potential critical quality attributes (pCQAs) including aggregation, average DAR, oxidation, and deamidation in 2 h. With multi-dimensional liquid chromatography, different classes of multi-product mAbs (cetuximab, panitumumab, rituximab) separation are also feasible in both elution modes with generic salt and p^H gradient CEX separation [29–33, 50, 51].

From the chiral analysis review, one study describes a 2D LC–MS approach that allows the simultaneous analysis of paracetamol and the two ketorolac enantiomers. Ketorolac is a non-steroidal anti-inflammatory drug (NSAID), which has a strong analgesic activity. It possesses a chiral center and is marketed as a racemic mixture of (+) R and (−) S enantiomers. The efficacy of combining paracetamol and ketorolac on numerous experimental pain models was evaluated in randomized placebo-controlled clinical trials in healthy human volunteers. As a result, an assay was needed to confirm the presence of these medicines in human plasma in order to characterize

Coupling	Analytes	Matrix	First dimension	Second dimension	2D-LC mode	References
Trace analysis (determination of drug traces into biological matrices)						
RPLC-RPLC	Tolterodine, Amperozide & metabolites	Biological matrices	Amide column Gradient elution Ammonium formate pH 3.6 /ACN F: 0.2 mL/min, T: 40°C	PFPP column Gradient elution Ammonium formate pH 3.6 /ACN F: 0.2 mL/min	Heart cut	[28]
RPLC-RPLC	API & related isomer API	Research compounds synthesized	Eclipse XDB-C18 gradient elution Water +0.1% FA/ACN F:1 mL/min, T: 25°C	XSelect CSH Phenyl-Hexyl gradient elution Water/ACN + 0.1% FA	Heart-cut	[18]
RPLC-RPLC	API & metabolites	Rats excreta	XTerra TM RP18 Gradient elution Water 0.1% FA/ACN F: 60 mL/min T: 40°C	Discovery C18 Gradient elution Water +10 mM ammonium acetate/ACN – T: 40°C	Heart-cut (Off-line)	[22]
RPLC-RPLC	Phenprocoumon and metabolites	Human plasma	Grom Sil ODS-3 CP C18 Gradient elution Ammonium acetate buffer pH 3.9/MeOH F: 200 µL/min T: 40°C	Chira Grom 2 Isocratic elution Water/ ACN + FAA pH 3.9, F: 150 µl/min T: 40°C	Heart cut	[20]
RPLC-RPLC	Terbutaline	Human plasma	Nucleosil-phenyl isocratic elution ammonium acetate pH 4.6 F: 0.2 mL/min T: 40°C	b-Cyclodextrin Isocratic elution ammonium acetate pH 6/MeOH F: 0.7 mL/min T: 40°C	Heart cut	[15]
RPLC-RPLC	API & impurities	Standard mixtures	Zorbax Eclipse XDB C18 Gradient elution potassium phosphate pH 6.5 in water/ACN F: 1.2 mL/min T: 40°C	Zorbax Eclipse XDB C18 Gradient elution Water +0.05% FA/ACN F: 1.2 mL/min T: 40°C	Heart cut	[21]
RPLC-RPLC	Duloxetine & impurities	Standard mixture	Zorbax Eclipse Plus C18 Isocratic elution Phosphate buffer/ACN/MeOH F: 1 mL/min	Zorbax Eclipse Plus C18 Isocratic elution Water/ACN + 0.1% formic acid F: As gradient	Heart cut	[29]

Coupling	Analytes	Matrix	First dimension	Second dimension	2D-LC mode	References
RPLC-RPLC	Taxanes	Standard mixtures/ extract samples	Eclipse Plus C18 Gradient elution Water/MeOH F: 0.06 mL/min T: 30°C	Eclipse Plus Phenyl-Hexyl Gradient elution Water/ACN + FA F: 4 mL/min T: 40°C	Comprehensive	[30]
IPC-RPLC	Phenylpropanolamine	Human plasma/urine	YMC ODS Isocratic elution phosphate buffer pH:3.5 + SBS/ ACNF: F;1 mL/min—T: 40°C	YMC ODS Isocratic elution phosphate buffer pH 3.5/CAN F: 1 mL/min—T: 40°C	Heart cut	[31]
NPLC–NPLC	Dexamethasone	Bovine tissue	Spherisorb phenyl Isocratic elution Water/acetic acid/2-propanol/ hexane F: 1.5 mL/ min T: 30°C	Spherisorb CN Isocratic elution Water/ acetic acid/2-propanol/ hexane F: 1.5 mL/min T: 30°C	Heart cut	[32]
RPLC-RPLC-RPLC	Mefenamic acid	Human serum	YMC ODS Isocratic elution Phosphate buffer pH:5/ACN F: 1 mL/min—T: 40°C	YMC ODS (1)—ODS 80 TM (2) Isocratic elution (1) Phosphate buffer pH 3.5/ACN (2) Phosphate buffer pH 6/ ACN F: 1 mL/min—T: 40°C	Heart cut	[33]
RPLC-RPLC	1,2,3,4-tetrahydro-1-naphthol (THN), hexobarbital (HXL)	Pharmaceutical drug development samples.	BEH C18, (150 mm × 2.1 mm × 1.7 μm), acetonitrile, 0.1% formic acid pH = 2 F: 0.15 mL/min T: 25°C	C18 (150 mm × 4.6 mm × column packed with 3 μm particles)), 15 mM NH_4TFA buffer acetonitrile, pH = 2 F: 0.25 and 1 mL/min T: 30°C	Heart cut	[34]
RPLC-RPLC	API & impurities	Standard mixture	XTerra RP18 column (4.6 × 150 mm, 3.5 m), mobile phase 10 mM potassium phosphate in water, pH 2.6 and ACN, F:1.0 mL/min, T: 30°C	C18 column (4.6 × 150 mm, 3.5 m) mobile phase 0.03% formic acid in water and CAN pH 4.0, F: 1.0 mL/min, T: 30°C	Heart cut	[21]
RPLC–RPLC	Vancomycin	Standard mixture/ Human plasma	RP ASTON C18 Isocratic elution Water/ ACN + ammonium acetate pH:3.8 F: 1 mL/min	ACR C18 Isocratic elution Water/ACN + ammonium acetate pH: 5.2 F: 1.2 mL/min	Heart cut	[35]

Coupling	Analytes	Matrix	First dimension	Second dimension	2D-LC mode	References
Chiral analysis						
RPLC-RPLC	Ketorolac & Paracetamol	Human Plasma	Discovery C18 Gradient elution Water +0.1% FA/ACN F: 0.2 mL/min	ChiralPak AD-RH Isocratic elution Water +0.1% FA/ACN F: 0.15 mL/min	Heart cut	[36]
IEX-RPLC	Propafenone	Human Plasma	Partisil SCX Gradient elution Water + Perchloric and phosphoric acids pH 2.4/ACN F: 0.7 mL/min	Chiralcel ODR Isocratic elution Water + Perchloric and phosphoric acids pH 2.4/ ACN F: 1.3 mL/min—T: 15°C	Heart cut	[37]
HILIC-HILIC	Salbutamol, Salmeterol, Atenolol	Urine samples	Kinetex HILIC Isocratic elution Acetate buffer pH: 6 /MeOH/ACN F: 0.4 mL/min T: 25°C	Isocratic elution Acetate buffer pH: 4 / MeOH F: 0.4 mL/min T: 25°C	Heart cut	[38]
NPLC-NPLC	Pimobendan & metabolites	Human plasma	Spheri-5 silicaIsocratic elution A: n-hexane/EtOH +0.1%Et2NHF: F:1 mL/min—T: 35°C	Chiralcel ODIsocratic elution A: n-hexane/EtOH +0.1%Et2NH F: 1 mL/min—T: 35°C	Heart cut	[39]
Capillary extraction column-NPLC	Antidepressant and Antiepileptic drugs	Urine samples	Graphene Oxide Capillary Column, (100 mm × 2.1 mm × 2.7 µm dp), H_2O/ACN (30%:70% 35%:65%), F: 0.20 mL/min, T: 40°C	C8,(200-mm length and 508-µm i.d.),H_2O/ACN(30%:70%, 35%:65%) , F: 0.05 mL/ min, T:150°C	Heart cut	[40]
sRPLC-SFC	Chiral active pharmaceutical ingredient elated Impurities	Complex pharmaceutical sample	Acquity BEH C18, Chiralpak IC column (50 mm × 2.1 mm 1.7 µm) MeOH/H_2O 98/2 (v/v) F: 0.2 mL/ min T: 40°C°C	Chiralpak IC column (150 mm × 4.6 mm; 3 µm), MeOH/H_2O 98/2 (v/v) F:2.5 mL/ min, T: 40°C	Multiple Heart-cutting	[41]
RPLC-RPLC	Hydroxywarfarins and warfarin	Human plasma	RP Acquity UPLC BEH Phenyl column (2.1 mm × 150 mm 1.7 µm) particle column, F:300 µL/min, T:60°C.	Chirobiotic V column [2.1 mm × 150 mm, 5 µm), room T: 21.6–22.4°C	Heart cut	[42]
RPLC-RPLC	Stereoisomers (from anti-HCV therapeutic)	Resarch compounds synthesized	Cortecs C18 Gradient elution Water +0,1% Phospohoric acid/ACN/MeOH F: 0,22 mL/min—T: 40°C	Teicoplanin Isocratic elution Water +0,1% Phosphoric acid/ACN	Heart cut	[43]

Coupling	Analytes	Matrix	First dimension	Second dimension	2D-LC mode	References
RPLC–RPLC	Leucovorin	Dog plasma	Resolvosil BSA-7 Isocratic elution Sodium phosphate in water pH 5.1 F: 1 mL/min T: 40°C	LiChrocart RP-18 gradient elution Sodium phosphate in water pH 5 /MeOH F: 1 mL/min T: ambiant	Heart cut	[44]
RPLC–NPLC	API & impurities	Research compounds synthesized	Symmetry Shield RP18 gradient elution Water +0.1% ortho-phosphoric acid/ACN F: 0.8 mL/min – T: 22°C	Chiralpak AD-H Isocratic elution hexane/isopropanol	Heart cut	[45]
RPLC–RPLC	Omeprazole	Human plasma	Restricted access media BSA octyl column Luna C8(100mm × 4.6 mm i.d., 10 μm) Water and ACN (30:70,v/v) F: 1 mL/min	Amylos tris(3,5 dimethylcarbamate) (150 mm × 4.6 mm i.d., 7 μm) Water and ACN (60:40,v/v) F: 0.5 mL/min	Direct Heart cut	[46]

RP: reversed phase, F: flow rate, T: temperature, FA: formic acid, BSA: bovine serum albumin, ACN: acetonitrile. BEH: ethylene bridge hybride, EtOH: ethanol, MeOH: Methano.

Table 3.
Multi-dimensional LC analysis in pharmaceutical products (trace and chiral analysis).

their pharmacokinetic profiles. The method consists of a gradient RPLC method on a C_{18} column coupled with a stereoselective, isocratic, RPLC method on polysaccharide chiral column, and the result explained in **Figure 1** [36].

Another instructive example is the combination of achiral and chiral separations in a single mLC–LC separation of warfarin and hydroxywarfarin isomers, where the first dimension resolves the majority of the analytes and the second dimension completes the job. The first dimension has a chiral stationary phase, whereas the second dimension has an achiral separation. This indicates that neither achiral nor chiral separation is sufficient to resolve the mixture. When the same separation modes are used in a multiple heart-cutting 2D-LC separation, the achiral reversed-phase separation used in the second dimension is highly complementary to the chiral separation used in the first dimension. These provide enough selectivity to quickly resolve those components of the sample that remains unresolved after passing through the 1D column [29, 34].

Another popular pharmaceutical application is comprehensive LC in stress-testing studies for different pharmaceutical products. The API's and intrinsic stability of other drug products should be assessed, and the degradation mechanisms should be disclosed. Stress testing, also called forced degradation study, helps to establish the intrinsic stability of the API. Stress testing includes a number of experiments, such as the effect of temperature, humidity, oxidation, photolysis, or hydrolysis at different p^H values, as outlined by the World Health Organization [33, 35, 36, 52].

The 1D-LC analysis of a strongly stressed (temperature, UV irradiation, and organic solvent) omeprazole tablet demonstrates this notion. One has the impression that in this study, the omeprazole peak is corresponding to only one product, the API, but recording a mass spectrum revealed that many solutes are co-eluting in the peak. Increasing resolution by utilizing longer columns or smaller particles is not a viable option because there is not enough peak capacity to resolve the solutes in the primary peak (zero resolution). Although choosing another stationary phase is an option, the risk of other peaks now overlapping is realistic. By far, the best solution is to combine different selectivities in an LC × LC combination even if the two mechanisms are comparable, for example, RPLC × RPLC but using different stationary and mobile phases [13, 35, 36].

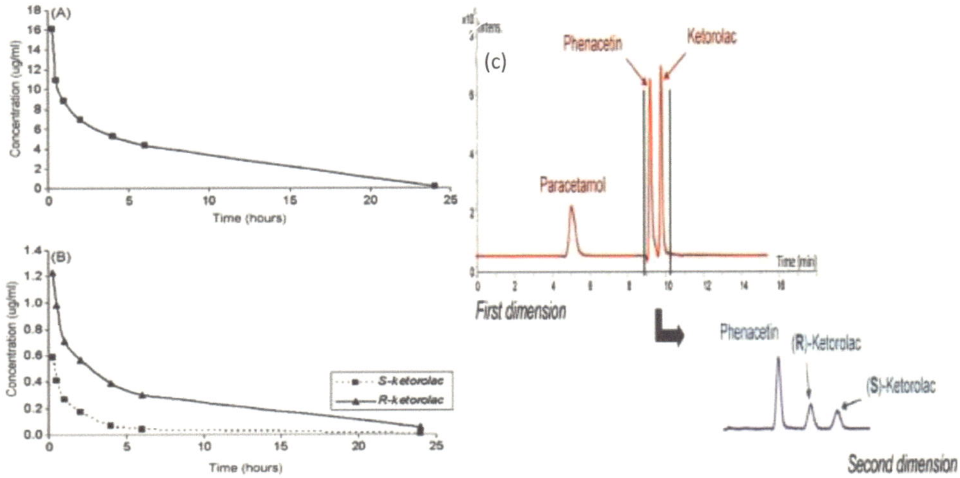

Figure 1.
Plasma concentration-time profile of (A) paracetamol (B) S and R-ketorolac, and (C) simultaneous analysis of both drugs after IV administration based on achiral-chiral 2D-LC.

Different mobile phase compositions with different selectivities were used in the first- and second-dimension columns. The study revealed more than 50 spots, with the omeprazole peak resolved in the second dimension into four distinct products, and its mass spectrum is also recorded. The conditions developed in this 2D-LC approach can be considered more or less generic for detection of impurities in APIs, at least when eluting under reversed phase LC conditions. In this robust and repeatable method, good finds were also achieved for analyses of metoclopramide, acetaminophen, diclofenac, ibuprofen, and lidocaine [32–34, 53–56].

4.2 Traditional medicines (TMs)

Regardless of the fact multi-dimensional chromatography is applicable to many traditional medicines, its application in traditional Chinese medicine is quite well researched. Traditional Chinese medicines are garnering attention in modern pharmaceutical institutes as a valuable resource for medication development. TCMs are extremely complex mixes having hundreds, if not thousands of components of various structures and concentrations, with only a few compounds responsible for specific pharmacological activity and/or toxicity. As a result, the use of latest analytical techniques is critical for the elucidation of the composition and quality control of TCMs [42, 43, 57, 58].

For the analysis of TCMs, a number of two-dimensional LCs have been designed with different column. Mostly, commercial columns with various separation techniques, such as strong cation-exchange chromatography or reverse phase liquid chromatography, have been coupled. Such coupled separation mode may reveal much more information on components due to a specific interaction with these biomacromolecules of traditional medicines. In most of the cases, the two columns have been combined off-line, which makes the process simpler. However, a few are truly comprehensive (LC × LC). *Ligusticum chuanxiong, Angelica sinensis, Swertia franchetiana,* and *Lonicera caprifolium* are some of the Chinese traditional medicines that are separated and identified by multi-dimensional liquid chromatography. Analysis of a traditional Chinese medicine using a two-dimensional cyano octadecyl silyl system with an eight port valves with two sample loops and UV APCI MS detection also gave over 52 components [40–42, 57, 58].

4.3 Lipidomics

The wide and complex lipid composition in biological samples requires MDLC methodologies to sufficiently separate the lipids prior to MS or other detector characterization. This lipid separation usually demonstrated with normal phase, reversed phase, and silver ion chromatography. The application of MDLC in complex field of lipidomics is becoming of increasing significance. One typical example of these lipidomics application is analysis of egg yolk by 2D high-performance liquid chromatography-mass spectrometry for phosphatidylcholine. Phosphatidylcholine is the main phospholipid present in egg yolk. For characterization of the fatty acids composition in phosphatidylcholines molecules in egg yolk, an off-line LC × LC-ESI-MS/MS (preparative C_{18} column) method using a triple quadrupole mass spectrometry was used. The study identified phosphatidylcholines, which contains unsaturated fatty acids from both omega-3 and 6 groups. Phosphatidyl ethanolamine, lysophosphatidylglycerol, sphingomyelin, lysophosphatidylethanolamine, lysophosphatidylcholine, and glycerophosphorylcholine were also among the 13 phospholipid fractions evaluated by this method [53, 54, 59–62].

4.4 Proteomics

MDLC has been used very successfully in proteomics for about two decades. These commonly comprehensive type separations, variant unique to peptide analysis, are known as multi-dimensional protein identification technology (MudPIT). Mass spectrometry has been widely used with as the most common detector. MDLC coupled with mass spectrometry is becoming increasingly important in proteome research owing to its high speed, high resolution, and high sensitivity. Recent proteomics technologies offer excellent separation and enormous data-gathering capabilities in the discovery of peptides and proteins, particularly disease-specific biomarkers in serum, plasma, urine, tissue, and other biological samples. The performance of multi-dimensional chromatography separation techniques compatible with MS, which are commonly used in proteomics applications, is summarized in the **Table 4** [45, 46, 53].

The first applications of multi-dimensional liquid chromatography for environmental research were performed at the end of the twentieth century. This application was known as column switching mode, which is widely employed for the separation and analysis of pesticides. Pol and coworkers published the first studies on the use of a comprehensive two-dimensional liquid chromatography system, termed LC–LC, for the analysis of environmental samples in 2006. This analysis showed acidic compounds present in atmospheric aerosols were separated using a multi-dimensional chromatographic system for the first time, proving the capabilities of LC–LC systems to separate intricate mixtures that would be too difficult for 1D-LC [67].

One of the environmental contaminants is endocrine-disrupting compounds (EDCs) containing chemicals and hormones having endocrine-disrupting activity. This compound becomes critical emerging contaminants due to their presence in environmental waters and worries about probable detrimental adverse effects to wild life and humans. A study done in Czech Republic was used multi heart-cutting two-dimensional liquid chromatography-atmospheric pressure photoionization-tandem mass spectrometry method for the determination of endocrine-disrupting compounds in water. Twenty real samples collected from the Loucka and the Svratka rivers were analyzed, and compounds were found in all Svratka samples (9.7–11.2 ng l^{-1} for -estradiol, 7.6–9.3 ng l^{-1} for estrone, and 24.6–38.7 ng l^{-1} for bisphenol A) [64, 65, 68].

3D-LC setup	Sample (protein amount)	Identified proteins	Identified unique peptides	MS time (hr)	Year	Reference
SCX-HILIC-RP	HeLa cells (500 μg)	3424	11,980	126	2013	[63]
SEC-HILIC-RP	Paracoccus denitrificans (8 mg)	2627	—	66	2015	[64]
RP-RP-RP	Jurkat cells (720 μg)	14,230	251,166	189	2016	[65]
SAX-RP-RP	HEK 293 T cells (30 μg)	8222	74,432	20.4	2017	[66]
SCX & SAX-RP RP	Human serum (−)	862	-	12	2018	[62]

SAX: Strong Anion Exchange, SCX: Strong Cation Exchange HILIC: Hydrophilic Interaction. Liquid Chromatography, HEK 293 T cells: Human embryonic kidney 293 cells.

Table 4.
Performance of various off-line 3D-LC systems in protein analysis.

Another novel multi-dimensional separation system based on online comprehensive two-dimensional liquid chromatography and hybrid high-resolution mass spectrometry has been developed for the qualitative screening analysis and characterization of wastewater sample. The core of the system is a consistently miniaturized two-dimensional liquid chromatography that makes the rapid second dimension compatible with mass spectrometry without the need for any flow split. Elevated temperature, ultrahigh pressure, and a superficially porous sub-3-μm stationary phase provide a fast second-dimensional separation and a sufficient sampling frequency without a first-dimensional flow stop. To seek data for a suspected target screening of a wastewater sample, 99 substances were added in the reference mix [62, 66].

Another novel method was developed utilizing LC × LC-ESI-TOF coupled MS for the determination of organic acids in atmospheric aerosols. They analyzed methanolic extracts of filters containing atmospheric aerosols and expected to find mainly polar organic compounds such as aliphatic, aromatic, and substituted carboxylic acids. Because of the combined knowledge of the elution pattern and the sensitive and accurate mass spectrum data, this innovative perspective liquid chromatography (LC-LC-TOF coupled MS) is proven to be an appropriate method for screening undiscovered acidic compounds [67, 69, 70].

4.5 Recent advances and future perspectives

Over the last decades, multi-dimensional liquid chromatography techniques have been extensively exploited with the latest significant improvements in terms of instrumental setup and availability of novel stationary phases. Nowadays, robust and full-featured instrumentations are available from most LC manufactures. Miniaturization and downscaling, switching valves as well as suitable software offer the possibility to adapt two-dimensional procedures and instrumentations will certainly continue in the future, and consequently, a significant rise of 2D LC systems is expected in several research fields [71].

Another new development in this area is Nano 2D-HPLC. 2D online nano-LC/MS was developed that substituted the inserted salt step gradient with an optimized semi-continuous pumped salt gradient, and also, 8-Isoprostaglandin F2α was measured from human urine. A microchip-based nano-HPLC was also used. So it will be the most applicable technique especially in the chiral resolution in the future [72–75].

Currently, a multi-dimensional liquid chromatography also gives different opportunity to those challenging areas for the analysis. Recently, major developments are seen and attracted significant interest toward complex in the analysis of small, more complex molecules and biological products. For example, biopharmaceuticals such as monoclonal antibody (mAbs), interferons/cytokines, and vaccines are recently analyzed by the advanced instrumental technology of multi-dimensional liquid chromatography [76].

5. Conclusion

Multi-dimensional liquid chromatography where a complex mixture is separated by facing different dimension is useful in several fields, especially in-depth analysis of proteome and lipidomics, environmental chemicals, food and pharmaceuticals industries. It boosts separation power, peak resolution, and reproducibility while increasing system complexity. The primary reason for multi-dimensional separations is that they provide a more effective and efficient method of generating high-peak capacities, and hence allowing for more comprehensive resolution of complicated mixtures. Separation and characterization of complicated mixtures are critical in a wide variety of fields that demand considerable separation power. Multi-dimensional separations envisaged that better understanding and application of multi-dimensional separations would open up opportunities for meaningful analyses of extremely complex samples and allowing the regular analysis of thousands of constituents from a single sample in a single run.

Author details

Esayas Tesfaye[1,2], Tadele Eticha[2], Ariaya Hymete[2] and Ayenew Ashenef[2,3*]

1 Department of Pharmacy, College of Health and Medical Sciences, Dilla University, Dilla, Ethiopia

2 Department of Pharmaceutical Chemistry, School of Pharmacy, College of Health Sciences, Addis Ababa University, Addis Ababa, Ethiopia

3 Center for Innovative Drug Development and Therapeutic Trials for Africa (CDT-Africa), College of Health Sciences, Addis Ababa University, Addis Ababa, Ethiopia

*Address all correspondence to: ayenew.ashenef@aau.edu.et

References

[1] Jandera P, Novotná K, Kolářová L, Fischer J. Phase system selectivity and peak capacity in liquid column chromatography - the impact on two-dimensional separations. Chromatographia. 2004;**60**:27-35. DOI: 10.1365/s10337-004-0354-1

[2] Fairchild JN. Multi-dimensional Liquid Chromatography Separations. Knoxville: University of Tennessee; 2010. pp. 1-185

[3] Dixon SP, Pitfield ID, Perrett D. Comprehensive multi-dimensional liquid chromatographic separation in biomedical and pharmaceutical analysis: A review. Biomedical Chromatography. 2006;**20**:508-529. DOI: 10.1002/bmc.672

[4] Frel FE, Frei RW. Two-dimensional column liquid chromatographic technique for resolution of complex mixtures. Journal of Chromatography A. 1978;**149**:561-569. DOI: 10.1016/S0021-9673(00)81011-0

[5] Dugo P, Cacciola F, Kumm T, Dugo G, Mondello L. Comprehensive multi-dimensional liquid chromatography: Theory and applications. Journal of Chromatography. A. 2008;**1184**:353-368. DOI: 10.1016/j.chroma.2007.06.074

[6] Giddings JC. Multi-dimensional chromatography-techniques and applications. AIChE. 1990;**36**(12):1933

[7] Emmett MR, Caprioli RM. Micro-electrospray mass spectrometry: Ultra-high-sensitivity analysis of peptides and proteins. Journal of the American Society for Mass Spectrometry. 1994;**5**:605-613. DOI: 10.1016/1044-0305(94)85001-1

[8] Giddings JC. Two dimensional separation concept and promise. Analytical Chemistry. 1984;**56**(12):1-13. DOI: 10.1021/ac00276a003

[9] O'Farrell PH. High resolution of proteins * two-dimensional electrophoresis. The Journal of Biological Chemistry. 1975;**250**(10):4007-4021. DOI: 10.1016/S0021-9258(19)41496-8

[10] Donato P, Cacciola F, Tranchida PQ, Dugo P, Mondello L. Mass specroscopy detection in comprehensive liquid chromatography: Basic concepts, instrunmental aspects, application and trends. Mass Spectrometry Reviews. 2012;**31**:523-559. DOI: 10.1002/mas.20353

[11] Pandohee J, Holland BJ, Li B, Tsuzuki T, Stevenson PG, Barnett NW, et al. Screening of cannabinoids in industrial-grade hemp using two-dimensional liquid chromatography coupled with acidic potassium permanganate chemiluminescence detection. Journal of Separation Science. 2015;**38**(12):2024-2032. DOI: 10.1002/jssc.201500088

[12] Olivova P, Daly AE, Gebler JC, Gilar M. Orthogonality of separation in two-dimensional liquid chromatography. Analytical Chemistry. 2005;**77**(19):6426-6434. DOI: 10.1021/ac050923i

[13] Ali I, Suhail M, Aboul-enein HY. Advances in chiral multi-dimensional liquid chromatography. Trends in Analytical Chemistry. 2019;**120**:1-19. DOI: 10.1016/j.trac.2019.115634

[14] Liu Z, Patterson DG, Lee ML. Geometric approach to factor analysis for the estimation of orthogonality and practical peak capacity in comprehensive two-dimensional separations. Analytical Chemistry. 1995;**67**(21):3840-3845. DOI: 10.1021/ac00117a004

[15] Edholm LE, Lindberg C, Paulson J, Walhagen A. Determination of drug enantiomers in biological samples by coupled column liquid chromatography and liquid chromatography-mass spectrometry. Journal of Chromatography A. 1988;**424**:61-72. DOI: 10.1016/s0378-4347(00)81076-2

[16] Schure MR. Limit of detection, dilution factors, and technique compatibility in multi-dimensional separations utilizing chromatography, capillary electrophoresis, and field-flow fractionation. Analytical Chemistry. 1999;**71**(8):1645-1657. DOI: 10.1021/ac981128q

[17] Murphy RE, Schure MR, Foley JP. Effect of sampling rate on resolution in comprehensive two-dimensional liquid chromatography. Analytical Chemistry. 1998;**70**(8):1585-1594. DOI: 10.1021/ac971184b

[18] Yang SH, Wang J, Zhang K. Validation of a two-dimensional liquid chromatography method for quality control testing of pharmaceutical materials. Journal of Chromatography. A. 2017;**1492**:89-97. DOI: 10.1016/j.chroma.2017.02.074

[19] Pirok BWJ, Stoll DR, Schoenmakers PJ. Recent developments in two dimensional liquid chromatography – Fundamental improvements for practical applications. Analytical Chemistry. 2018:24-68. DOI: 10.1021/acs.analchem.8b04841

[20] Kammerer B, Kahlich R, Ufer M, Gleiter CH, Laufer S. Achiral – Chiral LC / LC – MS / MS coupling for determination of chiral discrimination effects in phenprocoumon metabolism. Analytical Biochemistry. 2005;**339**: 297-309. DOI: 10.1016/j.ab.2005.01.010

[21] Zhang K, Li Y, Tsang M, Chetwyn NP. Analysis of pharmaceutical impurities using multi-heartcutting 2D LC coupled with UV-charged aerosol MS detection. Journal of Separation Science. 2013;**36**:2986-2992. DOI: 10.1002/jssc.201300493

[22] Kiffe M, Graf D, Trunzer M. Two-dimensional liquid chromatography / mass spectrometry set-up for structural elucidation of metabolites in complex biological matrices. Rapid Communications in Mass Spectrometry. 2007;**21**:961-970. DOI: 10.1002/rcm

[23] Liu Z, Li Y, Jiang S, Chen L. The combinations of separation modes in multi-dimensional high performance liquid chromatography(MDHPLC). Se Pu. 1997;**15**(6):490-493

[24] François I, Paccou A, Davy PW. Applications of Multi-dimensional Liquid Chromatography in the Pharmaceutical Industry. Waters, The Science of What's Possible: Milford, Massachusetts; 1731

[25] Washburn MP, Wolters D, Yates JR. Large-scale analysis of the yeast proteome by multi-dimensional protein identification technology. Nature Biotechnology. 2001;**19**(3):242-247. DOI: 10.1038/85686

[26] Leonhardt J, Teutenberg T, Tuerk J, Schl P. Analytical methods two-dimensional liquid chromatographic approaches coupled to high resolution mass spectrometry for the analysis of complex samples. Analytical Methods. 2015;**7**:7697-7706. DOI: 10.1039/c5ay01143d

[27] Sheldon EM. Development of a LC-LC-MS complete heart-cut approach for the characterization of pharmaceutical compounds using standard instrumentation. Journal of Pharmaceutical and Biomedical Analysis.

2003;**31**(6):1153-1166. DOI: 10.1016/S0731-7085(03)00017-7

[28] Jayamanne M, Granelli I, Tjernberg A, Edlund P. Development of a two-dimensional liquid chromatography system for isolation of drug metabolites. Journal of Pharmaceutical and Biomedical Analysis. 2010;**51**:649-657. DOI: 10.1016/j.jpba.2009.09.007

[29] Joseph S. Heart-cut 2D-LC/MS approach for pharmaceutical impurity identifi cation using an Agilent 6540 Q-TOF LC/MS. System. 2012:1-12

[30] Vanhoenacker G, David F, Sandra K, Sandra P, Reichenbach SE. Determination of Taxanes in Taxus sp. with the Agilent 1290 Infinity 2D-LC Solution. Santa Clara, California, United States: Agilent Technologies; 2014. pp. 1-10

[31] Yamashita K, Motohashi M, Yashiki T. High-performance liquid chromatographic determination of phenylpropanolamine in human plasma and urine , using column switching combined with ion-pair chromatography. Journal of Chromatography A. 1990; **527**:103-114. 10.1016/s0378-4347(00)82087-3

[32] Mclaughlin L, Henion JD. Determination of dexamethasone in bovine tissues performance liquid chromatography and capillary gas chromatography-mass spectrometry. Journal of Chromatography. 1999;**529**(1990):1-19. DOI: 10.1016/S0378-4347(00)83803-7

[33] Yamashita K, Motohashi M, Yashiki T. Column-switching techniques for high-performance liquid chromatography of ibuprofen and mefenamic acid in human serum with short-wavelength ultraviolet detection. Journal of Chromatography. 1991;**570**:329-338. DOI: 10.1016/0378-4347(91)80536-l

[34] Suryakant R, Chen K, Boon J, Hellings M, Wicht K, Lynen F. Development of an achiral-chiral 2-dimensional heart-cutting platform for enhanced pharmaceutical impurity analysis. Journal of Chromatography. A. 1628;**2020**:461425. DOI: 10.1016/j.chroma.2020.461425

[35] Li X, Wang F, Xu B, Yu X, Yang Y, Zhang L, et al. Determination of the free and total concentrations of vancomycin by two-dimensional liquid chromatography and its application in elderly patients. Journal of Chromatography B. 2014;**969**:181-189. DOI: 10.1016/j.jchromb.2014.08.002

[36] Ing-lorenzini KR, Desmeules JA, Besson M, Veuthey J, Dayer P, Daali Y. Two-dimensional liquid chromatography– Ion trap mass spectrometry for the simultaneous determination of ketorolac enantiomers and paracetamol in human plasma application to a pharmacokinetic study. Journal of Chromatography. A. 2009;**1216**:3851-3856. DOI: 10.1016/j.chroma.2009.02.071

[37] Lamprecht G, Stoschitzky K. Enantioselective analysis of R- and S-propafenone in plasma by HPLC applying column switching and liquid – Liquid extraction. Journal of Chromatography. 2009;**877**:3489-3494. DOI: 10.1016/j.jchromb.2009.08.024

[38] Rosales-conrado YN, Leo E. Chiral determination of salbutamol, salmeterol and atenolol by two-dimensional LC – LC: Application to urine samples. Chromatographia. 2012;**75**:1365-1375. DOI: 10.1007/s10337-012-2353-y

[39] Chu KM, Shieh SM, Wu SH, Yoa-Pu HO. Enantiomeric separation of a cardiotonic agent pimobendan and

its major active metabolite, UD-CG 212 BS, by coupled achiral-chiral normal-phase high-performance liquid chromatography. Journal of Chromatographic Science. 1992;**30**:171-176. DOI: 10.1093/chromsci/30.5.171

[40] Vasconcelos E, Maciel S, Ana L, To D, Alves S. Multi-dimensional liquid chromatography employing a graphene oxide capillary column as the first dimension: determination of antidepressant and antiepileptic drugs in urine. Molecules. 2020;**25**(1092):1-14. DOI: 10.3390%2Fmolecules25051092

[41] Iguiniz AM, Corbel E, Roques N, Heinisch S. On-line coupling of achiral reversed phase liquid chromatography and chiral supercritical fluid chromatography for the analysis of pharmaceutical compounds. Journal of Pharmaceutical and Biomedical Analysis. 2018;**159**:1-21. DOI: 10.1016/j.jpba.2018.06.058

[42] Jones DR, Boysen G, Miller GP. Novel multi-mode ultra performance liquid chromatography-tandem mass spectrometry assay for profiling enantiomeric hydroxywarfarins and warfarin in human plasma. Journal of Chromatography B. 2011;**879**(15-16):1056-1062. DOI: 10.1016/j.jchromb.2011.03.022

[43] Gomez-Bengoa E, Helm M, Plant A. Ultrafast chiral chromatography as the second dimension in two-dimensional liquid chromatography experiments. Analytical Chemistry. 2017;**89**:3545-3553. DOI: 10.1021/acs.analchem.6b04834

[44] Silan L, Jadaud P, Whitfield LR, Wainer IW. Determination of low levels of the stereoisomers of leucovorin and 5-methyltetrahydrofolate in plasma using a coupled chiral-achiral high-performance liquid chromatographic system with post-chiral column peak compression. Journal of Chromatography. B, Biomedical Sciences and Applications. 1990;**532**(C):227-236. DOI: 10.1016/S0378-4347(00)83774-3

[45] Lee C, Zang J, Cuff J, Mcgachy N, Natishan TK, Welch CJ, et al. Application of heart-cutting 2D-LC for the determination of peak purity for a chiral pharmaceutical compound by HPLC. Chromatographia. 2013;**76**:5-11. DOI: 10.1007/s10337-012-2367-5

[46] Cass QB, Lima VV, Oliveira RV, Cassiano NM, Degani ALG, Pedrazzoli J. Enantiomeric determination of the plasma levels of omeprazole by direct plasma injection using high-performance liquid chromatography with achiral-chiral column-switching. Journal of Chromatography B. 2003;**798**:275-281. DOI: 10.1016/j.jchromb.2003.09.053

[47] Mahadik M, Bhusari V, Kulkarni M, Dhaneshwar S. LC – UV and LC – MS evaluation of stress degradation behaviour of tenatoprazole. Journal of Pharmaceutical and Biomedical Analysis. 2009;**50**:787-793. DOI: 10.1016/j.jpba.2009.06.026

[48] Perkel J. Biopharmaceutical chemistry research & development: New studies of biologic drugs, therapies, and biomedical analysis. Nature. 2014;**524**:29-31

[49] Aggarwal S. What's fueling the biotech engine-2009-2010. Nature Biotechnology. 2010;**28**(11):1165-1171. DOI: 10.1038/nbt1110-1165

[50] Cohen SARS. Multi-dimensional Liquid Chromatography Theory and Applications in Industrial. Hoboken, New Jersey, United States: John Wiley & Sons; 2007. pp. 1-19

[51] Fekete S, Beck A, Fekete J, Guillarme D. Method development for

the separation of monoclonal antibody charge variants in cation exchange chromatography, part I: Salt gradient approach. Journal of Pharmaceutical and Biomedical Analysis. 2015;**102**:33-44. DOI: 10.1016/j.jpba.2014.08.035

[52] Blessy M, Patel RD, Prajapati PN, Agrawal YK. Development of forced degradation and stability indicating studies of drugs - a review. Journal of Pharmaceutical Analysis. 2014;**4**(3):159-165. DOI: 10.1016/j.jpha.2013.09.003

[53] Jones DR, Kim SY, Guderyon M, Yun CH, Moran JH, Miller G. Hydroxywarfarin metabolites potently inhibit CYP2C9 metabolism of S-Warfarin. Chemical Research in Toxicology. 2016;**176**(1):100-106. DOI: 10.1021%2Ftx1000283

[54] Regalado EL, Schariter JA, Welch CJ. Investigation of two-dimensional high performance liquid chromatography approaches for reversed phase resolution of warfarin and hydroxywarfarin isomers. Journal of Chromatography. A. 2014;**1363**:200-206. DOI: 10.1016/j.chroma.2014.08.025

[55] Seshadri RK, Raghavaraju TV, IEC. A single gradient stability-indicating reversed-phase LC method for the estimation of impurities in omeprazole and domperidone capsule. Scientia Pharmaceutica. 2012;**81**:1-22. DOI: 10.3797/scipharm.1209-12

[56] Shankar G, Borkar RM, Udutha S, Kanakaraju M, Sai Charan G, Misra S, et al. Identification and structural characterization of the stress degradation products of omeprazole using Q-TOF-LC-ESI-MS/MS and NMR experiments: Evaluation of the toxicity of the degradation products. New Journal of Chemistry. 2019;**43**(19):7294-7306. DOI: 10.1039/c9nj00932a

[57] Shen Y, Van Beek TA, Zuilhof H, Chen B. Applications of liquid chromatography in the quality control of traditional chinese medicines, an overview. In: Liquid Chromatography: Applications. Amsterdam, Netherlands: Elsevier Inc.; 2013. pp. 519-540. DOI: 10.1016/B978-0-12-415806-1.00019-X

[58] Zhang J, Jin Y, Liu Y, Xiao Y, Feng J, Xue X, et al. Purification of alkaloids from corydalis yanhusuo W. T. Wang using preparative 2-D HPLC. Journal of Separation Science. 2009;**32**(9):1401-1406. DOI: 10.1002/jssc.200800729

[59] Gładkowski W, Kiełbowicz G, Chojnacka A, Gil M, Trziszka T, Dobrzański Z, et al. Fatty acid composition of egg yolk phospholipid fractions following feed supplementation of Lohmann Brown hens with humic-fat preparations. Food Chemistry. 2011;**126**(3):1013-1018. DOI: 10.1016/j.foodchem.2010.11.112

[60] Walczak J, Bocian S. Two-dimensional high performance liquid chromatography-mass spectrometry for phosphatidylcholine analysis in egg yolk. Food Analytical Methods. 2015;**8**:661-667. DOI: 10.1007/s12161-014-9942-3

[61] Mondello L, Tranchida PQ, Stanek V, Jandera P, Dugo G, Dugo P. Silver-ion reversed-phase comprehensive two-dimensional liquid chromatography combined with mass spectrometric detection in lipidic food analysis. Journal of Chromatography. A. 2005;**1086**(1-2):91-98. DOI: 10.1016/j.chroma.2005.06.017

[62] Xue L, Lin L, Zhou W, Chen W, Tang J, Sun X, et al. Mixed-mode ion exchange-based integrated proteomics technology for fast and deep plasma proteome profiling. Journal of

Chromatography. A. 2018;**1564**:76-84. DOI: 10.1016/j.chroma.2018.06.020

[63] Zhou H, Palma S Di, Preisinger C, Polat AN, Heck AJR, Mohammed S. Towards a comprehensive characterization of a human cancer cell phosphoproteome. Journal of Proteome Research 2012;1-29. DOI: 10.1021/pr300630k

[64] Pernikářová V, Sedláček V, Potěšil D, Iva Struhárová Z, Pavel Bouchal IK. Proteomic responses to amethyl viologen-induced oxidative stress in the wild type and 2 FerB mutant strains of Paracoccus denitrificans. Journal of Proteomics. 2015;**125**:1-8. DOI: 10.1016/j.jprot.2015.05.002

[65] Spicer V, Ezzati P, Neustaeter H, Beavis RC, Wilkins JA, Krokhin OV. 3D HPLC-MS with reversed-phase separation functionality in all three dimensions for large-scale bottom-up proteomics and peptide retention data collection. American Journal of Analytical Chemistry. 2016;**88**(5):1-25. DOI: 10.1021/acs.analchem.5b04567

[66] Yang P, Tian R. 3D-SISPROT: A simple and integrated spintip-based protein digestion and three-dimensional peptide fractionation technology for deep proteome profilin. Journal of Chromatography. A. 2017:1-8. DOI: 10.1016/j.chroma.2017.01.033

[67] Hogendoorn EA, Goewie CE, Zoonen P Van. Application of HPLC column-switching in pesticide residue analysis. Journal of Analytical Chemistry 1991;**1**(339):348-356. DOI: 10.1007/BF00322347

[68] Matějíček D. On-line two-dimensional liquid chromatography-tandem mass spectrometric determination of estrogens in sediments. Journal of Chromatography. A. 2011;**1218**(16):2292-2300. DOI: 10.1016/j.chroma.2011.02.041

[69] Matˇ D. Multi heart-cutting two-dimensional liquid chromatography – Atmospheric pressure photoionization-tandem mass spectrometry method for the determination of endocrine disrupting compounds in water. Journal of Chromatography. A. 2012;**1231**:52-58. DOI: 10.1016/j.chroma.2012.02.006

[70] Ng NL, Canagaratna MR, Zhang Q, Jimenez JL, Tian J, Ulbrich IM, et al. Organic aerosol components observed in northern hemispheric datasets from aerosol mass spectrometry. Atmospheric Chemistry and Physics. 2007;**978**:1-10. DOI: 10.5194/acp-10-4625-2010

[71] Cacciola F, Mangraviti D, Mondello L, Dugo P. Hyphenations of 2D Capillary-based LC With Mass Spectrometry. Amsterdam, Netherlands: Elsevier Inc.; 2020. pp. 369-412. DOI: 10.1016/B978-0-12-809638-3.00010-7

[72] Gomez-Bengoa E, Helm M. Online and splitless NanoLC × CapillaryLC with quadrupole/time-of-flight mass spectrometric detection for comprehensive screening analysis of complex samples. Analytical Chemistry. 2014;**85**(21):10083-10090

[73] Hohnov B, Jussila M, Hy T. Comprehensive two-dimensional liquid chromatography – Time-of-flight mass spectrometry in the analysis of acidic compounds in atmospheric aerosols. Journal of Chromatography. A. 2006;**1130**:64-71. DOI: 10.1016/j.chroma.2006.04.050

[74] Ali I, Suhail M, Aboul HY, Tatiana E. Recent trends in chiral separations by 2D - HPLC. Chromatographia. 2021;**84**(6):535-548. DOI: 10.1007/s10337-021-04030-1

[75] Bai HY, Lin SL, Chung YT, Liu TY, Chan SA, Fuh MR. Quantitative determination of 8-isoprostaglandin F

2α in human urine using microfluidic chip-based nano-liquid chromatography with on-chip sample enrichment and tandem mass spectrometry. Journal of Chromatography. A. 2011;**1218**(15):2085-2090. DOI: 10.1016/j.chroma.2010.10.091

[76] Samanidou VF. Pharmaceutica analytica acta two-dimensional liquid chromatography (2D-LC) for biopharmaceuticals. Pharmaceutica Analytica Acta. 2018;**9**(5):9-10. DOI: 10.4172/2153-2435.1000e192

Chapter 4

Ionic Liquids in Liquid Chromatography

Victor David and Serban C. Moldoveanu

Abstract

Ionic liquids (ILs) are salts of organic cations that are present in liquid state. They can be used as alternative to organic solvents for various analytical processes such as extracting solvents in sample preparation, or as mobile phase or components of the mobile phase in high performance liquid chromatography (HPLC). Also they can be used as stationary phase in gas chromatography (GC), or attached to a solid support as stationary phase in HPLC. Ils are typically more environmentally-friendly solvents than the classic organic solvents having low volatility, flammability and toxicity. The chapter presents various applications of ILs in liquid chromatography.

Keywords: ionic liquids, HPLC, grafted ionic liquids, imidazolium salts, pyridinium salts

1. Introduction

Ionic liquids (ILs) are the salts of organic cations in the liquid state. The most common cations forming ionic liquids are pyridinium, piperidinium, and imidazolium. In addition, ammonium, guanidinium, sulfonium, phosphonium, or oxonium having various alkyl chains can also form ionic liquids. The counteranions can be small organic or inorganic ions, such as halogen-based anions (Cl^-, Br^-, BF_4^-, and $AlCl_4^-$), bis(trifluoromethanesulfonyl)imide, $[(CF_3SO_2)_2N]^-$, trifluoroacetate $[CF_3COO^-]$, or trifluoromethanesulfonate $[CF_3SO_3^-]$ [1]. Examples of some common organic cations and anions in the composition of ionic liquids are given in **Figure 1**.

ILs are currently used as an alternative to the organic solvents for various analytical processes [1], considering that they can be more environmentally friendly solvents than the classic organic solvents, due to their low volatility, flammability, and toxicity. The liquid property of these ionic compounds can be explained by the role of the short-range interactions and ion packing, combined with their long molecular structure [2]. Owing to the ability of ionic liquids to solvate the compounds of widely varying polarity [3], they are more and more utilized in separation processes. Examples of analytical separations based on ILs are liquid-liquid extraction (single drop microextraction, dispersive liquid-liquid microextraction, and hollow fiber membrane liquid phase) [4–6], solid-phase extraction and microextraction [7], gas chromatography (stationary phases based on ILs) [8, 9], electrophoresis [10, 11], and high-performance liquid chromatography (HPLC), discussed further in this chapter.

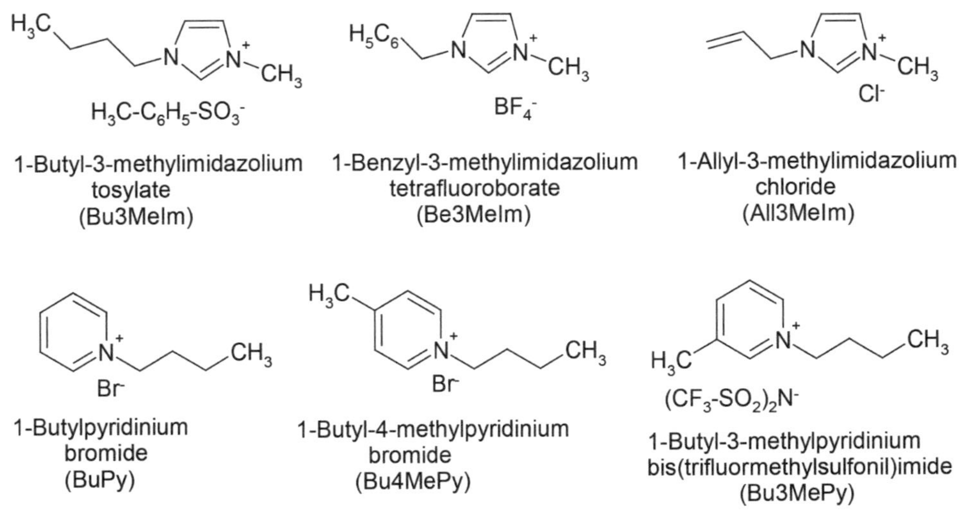

Figure 1.
Chemical structures, abbreviations, and names of several common ionic liquids.

The properties of ILs in the separation process depend on their interaction properties with surrounding molecules or surfaces, ion exchanging, H-bonding, ion-dipole, dipole-dipole, π-π interactions, hydrophobic, and hydrophilic interactions [12]. The interaction properties of ILs depend on the nature of both their anion and cation. The H-bonding interactions with water, by means of anion, determine the hydrophobic or hydrophilic character of IL. For example, for the IL based on the cation 1-butyl-3-methylimidazolium [C4mim], its hydrophobicity depends on the anion size: for anions, such as NO_3^-, or BF_4^-, the interactions with water molecules are strong, and consequently, the IL becomes hydrophilic, while for anions PF_6^- or bis(trifluoromethanesulfonyl)-imide [$(CF_3\text{-}SO_2)_2N^-$ or NTf2], the IL assembly becomes hydrophobic [13]. The nonpolar (hydrophobic) parts of ILs interact preferentially with nonpolar compounds by van der Waals forces, while the interface between the polar and nonpolar parts of ILs can interact by dipole-dipole with polar compounds, such as halogenated hydrocarbons or acetone [14]. Besides solubility in different solvents, the transport properties (viscosity, diffusion, and conductivity) can be described using theories on the structure–property correlations [15].

2. Stationary phases with attached structures of ionic liquids

There are a large variety of stationary phases containing ILs as an active functional group. Owing to their complex structure, these stationary phases are used in different separation mechanisms, including reversed-phase (RP), normal-phase (NP), hydrophilic interaction liquid chromatography (HILIC), ion-exchange chromatography (IEC), or combination of these mechanisms, known as mixed-mode separations. Generally, the stationary phases in HPLC having in their structure moieties of ionic liquids are synthetized having silica gel (type B) or derivatized silica gel as starting materials. Stationary phase can also be obtained by the polymerization of ionic liquids or by monolithic technologies. The most common types of IL-based stationary phases are bonded phases with a single-cation ionic liquid, obtained as spherical particles [16]. Ionic liquid stationary phases with multication ionic liquids and mixed functionalities between ionic liquids and hydrophobic or polar groups are also known in liquid chromatography [16].

The simplest method to introduce an IL moiety to the silica support is based on the two-step synthesis route illustrated in **Figure 2**. The first reaction is generally used to obtain chloropropyl silica, as intermediate material, from silica gel and 3-chloropropyltrimethoxysilane. The second reaction, between substituted imidazole or pyridine and chloropropyl silica gel, takes place in toluene under reflux [17, 18]. Longer linker than propyl is also used in attaching imidazole to the silica surface, such as octyl [19]. Imidazolium-based ILs with various functionalities are produced in order to obtain stationary phases with different retention properties: methyl, benzyl, and butylsulphonic [19].

Direct reaction between silica gel and imidazole-based derivatization reagent is also possible. Thus, the reaction between 3-bromopropyl-trimethoxysilane and 1-alkylimidazole can be used to attach IL to the silica gel surface [20, 21], as schematically shown in **Figure 3**.

Thiopropyl silica gel can also be used for obtaining a stationary phase with a single-cation IL based on a synthesis similarly to the schema presented in **Figure 2**,

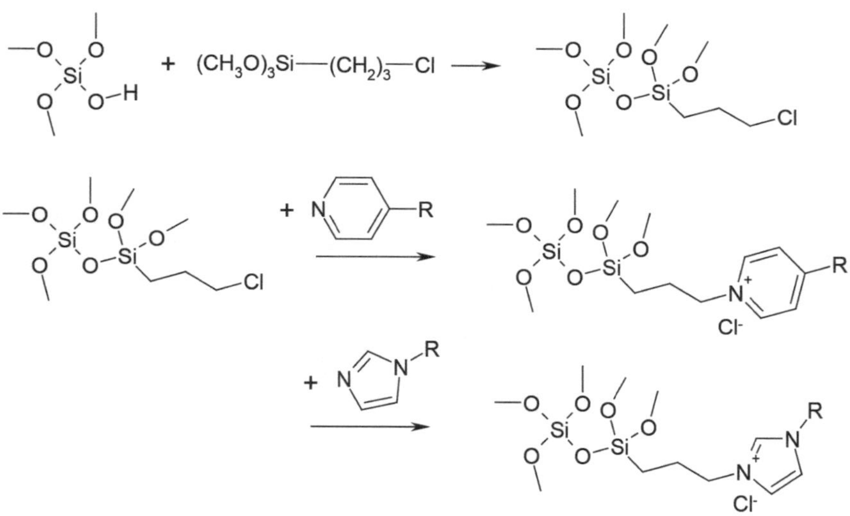

Figure 2.
The synthesis route for attaching imidazolium and pyridinium IL structure to the surface of silica gel (R can be C8, C10, C18, benzyl, or antranyl).

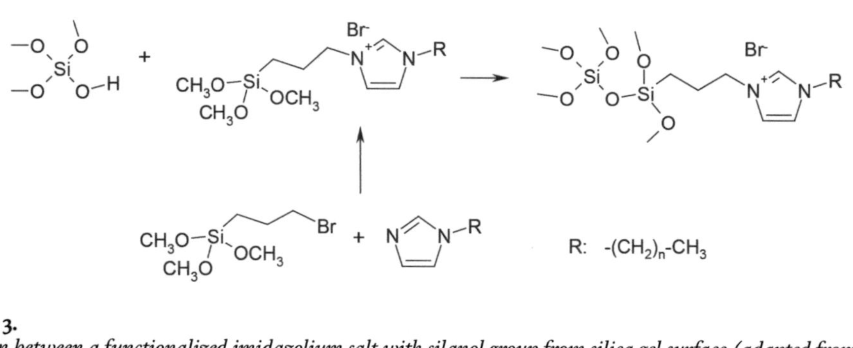

Figure 3.
Reaction between a functionalized imidazolium salt with silanol group from silica gel surface (adapted from [21]).

but based on addition reaction of thio group to the allyl radical of imidazolium-based IL (known also as "thiol-ene" click reaction). The structural feature of this stationary phase [22], used in the HILIC separation mechanism, is given in **Figure 4**.

Monolithic stationary phases are also reported in the literature [24], also known as supported poly(ionic liquid)s (SPILs) [25]. The most convenient method to prepare these stationary phases is based on monolithic support obtained from glycidyl methacrylate as monomer, using ethylene glycol dimethacrylate as crosslinker [26, 27]. Usually, the polymerization reaction is thermally initiated 2,2′-azobis-isobutyronitrile) by ultraviolet (UV) or γ radiation in a porogenic solvent (generally, they are the mixtures of cyclohexanol and dodecanol) [28]. The crosslinked structure contains epoxy groups, which are very reactive and can be used to react with imidazolium-based compound leading to more complex structures containing IL moiety at its surface [29]. An example of obtaining such monoliths is given in **Figure 5**.

Another possibility of obtaining polymers with ILs grafted to its surface is described in **Figure 6**. In this procedure, the monomer contains the imidazolium IL, which is then grafted to the surface of silica gel having thiopropyl as linker [30].

Based on this route, two types of stationary phases were obtained bearing two different anions: 1-[2-alkyl(mercaptopropyl)]-3-octadecyl-imidazolium bromide and 1-[2-alkyl(mercaptopropyl)]-3-octadecyl-imidazolium p-dimethylaminoazobenzenesulfonate [31]. Their chemical structures are shown in **Figure 7**.

The copolymerization of anionic (vinylsulphonic acid) and cationic (vinylimidazolium substituted with long alkyl chain) monomers that were grafted on thiopropyl silica gel was used to obtain the stationary phase with mixed-mode characters, hydrophilic due to the IL moiety and reversed-phase due to the long alkyl chain attached to the imidazolium ring [32]. A structure of such material, where the radical *R* is typically a long alkyl chain (C18), is shown in **Figure 8**. These materials exhibited an increased stability and improved peak shape of separated compounds, such as polycyclic aromatic hydrocarbons, bases, and flavonoids.

A novel facile and efficient immobilization strategy was introduced on the basis of the reaction between an isocyanate derivative of IL and aminopropyl silica gel

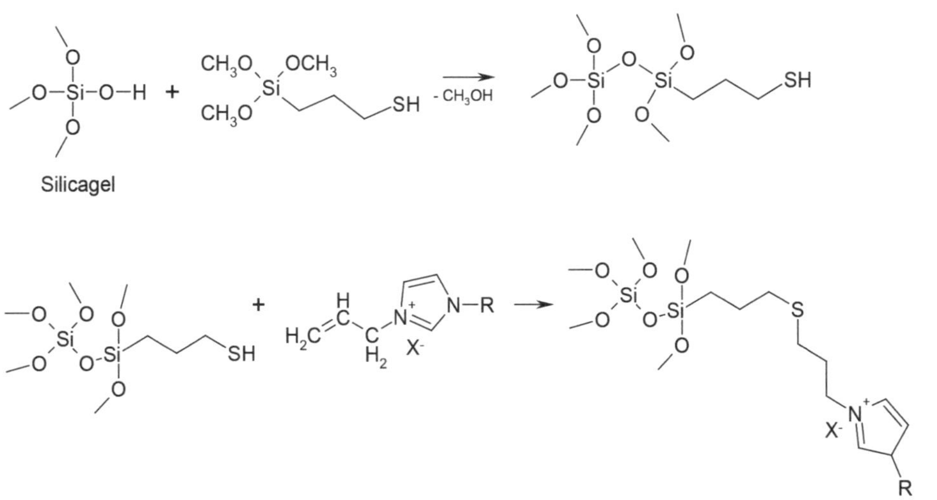

Figure 4.
Stationary phase synthetized from thiopropyl silica gel (obtained from silica gel in the first reaction) by attaching an imidazolium-based IL (adapted from [23]).

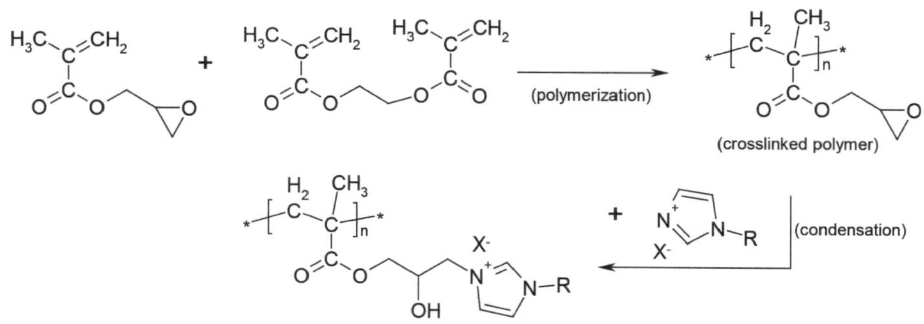

Figure 5.
Synthesis route to obtain a monolithic stationary phase based on glycidyl methacrylate and imidazolium moiety on its surface.

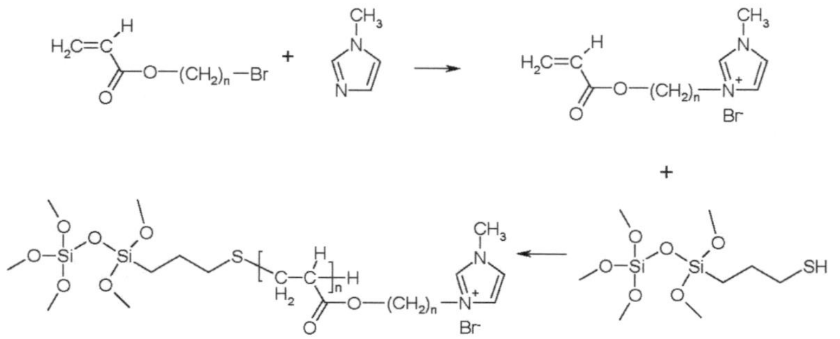

Figure 6.
Polymeric structure containing imidazolium IL grafted to the surface of thiopropyl silica surface (adapted from [30]).

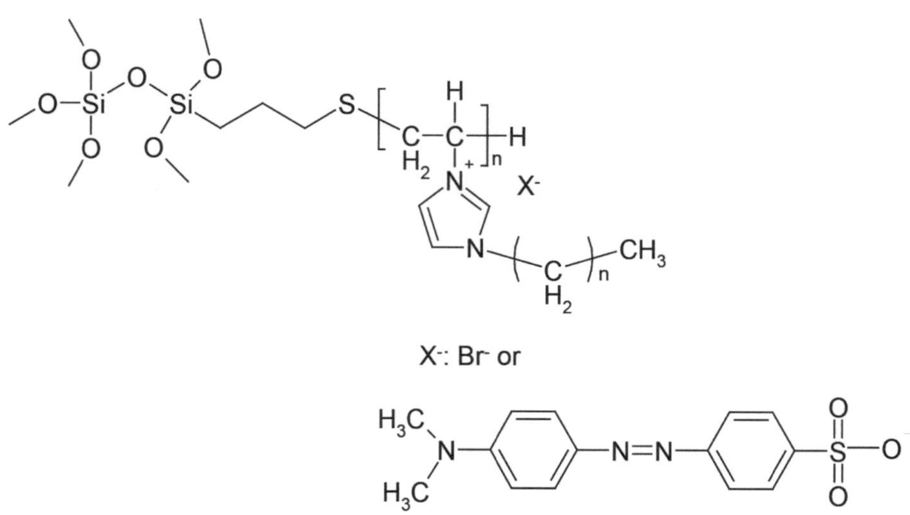

Figure 7.
Structure of 1-[2-alkyl(mercaptopropyl)]-3-octadecyl-imidazolium bonded to silica gel obtained by polymerization.

as mesoporous silica spheres. The result of this reaction is a multifunctional silica material used as HPLC stationary phase in the mixed-mode mechanism, with urea polar functionality and imidazolium salt bearing aliphatic chains of different lengths

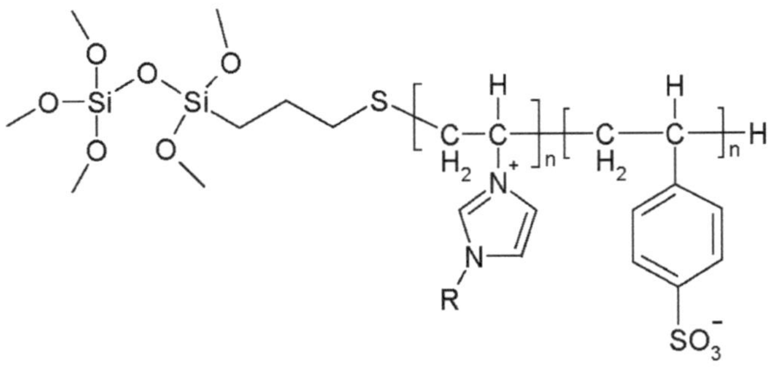

Figure 8.
Chemical structure of copolymerized vinylsulphonic acid and R-substituted vinylimidazolium grafted to the mercaptopropyl silica gel (adapted from [32]).

or aromatic groups [33]. The main derivatization reaction of aminopropyl silica gel with an isocyanate reagent bearing imidazolium salt is presented in **Figure 9**.

There have been synthetized stationary phases containing three ILs attached to mercaptopropyl silica gel with different anions [34]. Two examples of stationary phases used in reversed-phase and hydrophilic separations are presented in **Figure 10**.

More complex structures containing IL moieties and used as stationary phases in HPLC are reported by the literature [16, 21]. An example of such a complex stationary phase is based on the use of graphene oxide, which plays the role of linker between ILs and the silica support, which is easy to make further chemical modification [35]. Besides silica gel, or organic monoliths used as stationary phase supports, there are attempts to use zirconia as support to improve their thermal stability [36]. Stationary phases with ILs with chiral properties are synthetized from cyclodextrin bonded on silica gel, where the positively charged imidazole structure provides electrostatic interactions with opposite charged groups from analytes. Thus, chiral ILs functionalized β-cyclodextrins (β-CDs) were synthetized by treating 6-tosyl-β-cyclodextrin with 1,2-dimethylimidazole, which were bonded to silica gel to obtain chiral stationary phases (CSPs) to be used in high-performance liquid chromatography. There are applications when they have been used in separating different chiral aromatic alcohol derivatives and racemic drugs in mobile phase based on acetonitrile [37].

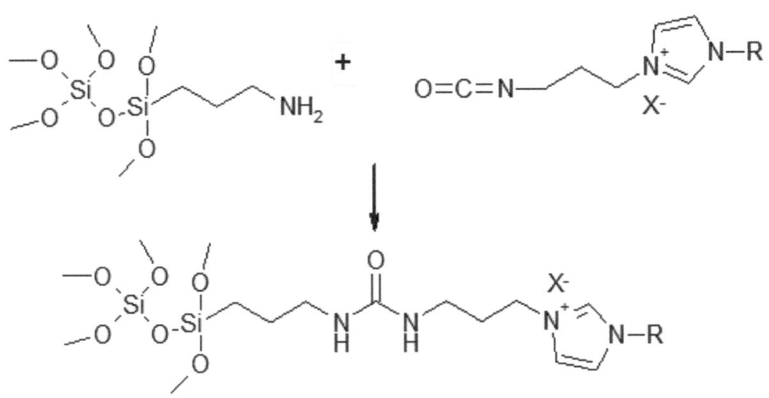

Figure 9.
Reaction between aminopropyl silica gel and isocyanate generating IL structure on silica gel surface (adapted from [33]).

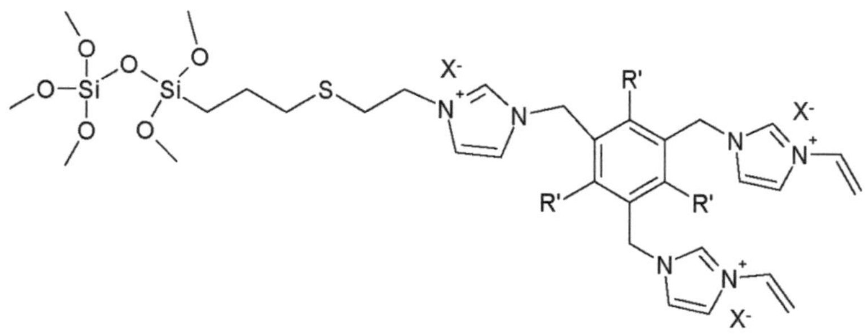

Figure 10.
Structure of tricationic ionic liquid attached to silica gel as HPLC stationary phases.

Stationary phases based on other IL moieties have been recently reported by the literature [38]. For example, phosphonium-based ionic stationary phase used in the mixed-mode mechanism was synthetized by polymerization from trioctyl(allyl) phosphonium bromide (denoted by [P888Allyl]Br, obtained at its turn from trioctylphosphine and allyl bromide) and mercaptopropyl silica gel microsphere, based on the principle of thiol-ene click reaction, described schematically in **Figure 4**. This stationary phase proved useful in separating the mixtures of uracil, uridine, adenosine, cytidine, and cytosine, various drugs, and mixtures of hydrophobic compounds (e.g., series from benzene to butylbenzene).

3. Ionic liquids as additives in mobile phase

The addition of ionic liquids in mobile phase increases its polarity, and this parameter can be useful in some separation mechanisms. On the other hand, the presence of these species in mobile phase can negatively influence the spectral domain used for UV-Vis detection. One of the main applications of these additives is to suppress the interference of residual silanols in the retention mechanism based on hydrophobic stationary phases.

The influence of residual silanols from the surface of reversed-phase stationary phases, based on derivatized silica on the retention of basic compounds is known to produce peak asymmetry and tailing [39]. A solution to improve the shape of peaks and eliminate peak tailing is the addition of ionic liquids in the composition of mobile phase, which have silanol suppressing properties [40]. This is based on the competition between the two adsorption equilibria, of ionic liquid cation and of basic analytes to the silanols, and thus, the basic analytes are less retained on these sites from the stationary phase surface. In practice, the concentration of ionic liquids is higher than that of analyte, and the analyte is practically excluded from the interaction with silanols. This effect has been observed in different applications, such as, for example, the separation of catecholoamines [41], β-lactam antibiotics [42], nucleotides [43], ephedrine derivatives [44], and fluoroquinolone [45], using imidazolium or pyridinium tetrafluoroborate ILs. However, the HPLC retention behavior of analytes in the presence of ILs in the mobile phase is influenced by both the cation and the anion due to their dissociation in aqueous medium [46].

The effect of ILs as additive on the retention of analytes depends on their polarity. The cation of IL can form ion pairs with the analytes as anions, and consequently, their

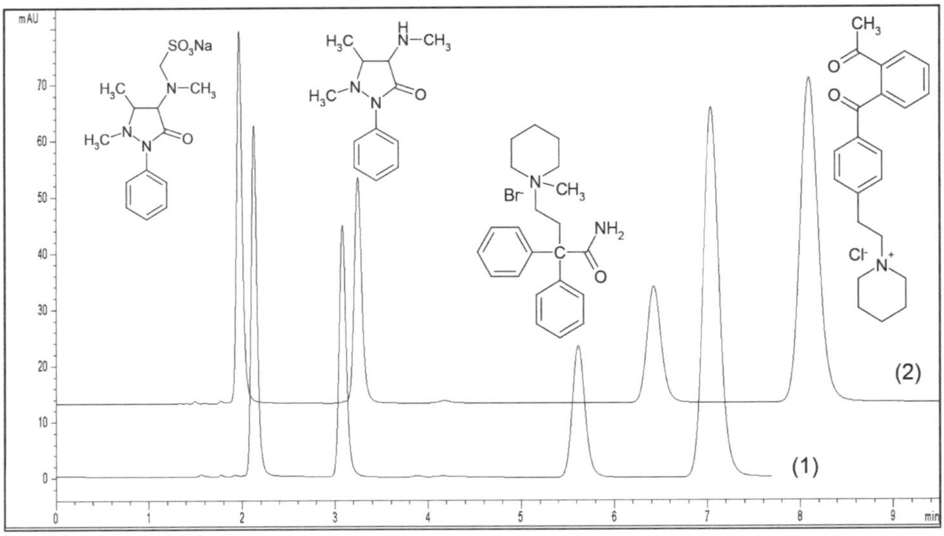

Figure 11.
Comparison between the retention of four pharmaceutical compounds (metamizole Na, metamizole impurity, fenpiverine bromide, and pitofenone hydrochloride, in the elution order), when the aqueous mobile phase contains: (1) both sodium hexane sulphonate and ionic liquid at 10 mM and (2) only 10 mM sodium hexane sulphonate (pH of the mobile phase is 3).

retention can be increased in reversed-phase HPLC. On the other hand, the retention of organic cations is decreased by the presence of IL cation. For a complex mixture, the retention order and separation selectivity can be modified by the presence of IL in the composition of mobile phase. One example is given in **Figure 11**, where the separation of four compounds from a pharmaceutical formulation by the ion-pairing mechanism with sodium hexane sulphonate is carried out in the presence and absence of 1-butyl-1-methyl-pyrrolidinium tetrafluoroborate [47]. Usually, the increase of IL concentration in the mobile phase leads to an increase of analyte retention [48].

4. Retention behavior of ionic liquids studied as analytes in HPLC mechanisms

Separation between ionic liquids as analytes or between them and other organic species is possible by HPLC due to the differentiation in interacting with the stationary phase. The injection of samples containing ionic liquids as analytes in organic solvents has been studied by several chromatographic mechanism discussed as follows. The main possible interactions of ionic liquids, exemplified for imidazolium-based ILs with various moieties from HPLC stationary phases leading to retention, are illustrated in **Figure 12**. These interactions depend on the type of the stationary phase and, in many cases, on the composition of the mobile phase [49].

The retention on the octadecyl silica surface in the absence of ion pairing agents has revealed strong interactions between imidazolium-based cations and stationary phase, which can be used in their separation from mixtures. For a series of alkyl-imidazolium ionic liquids, the dependence of the retention factor (k), expressed as logarithm, and the number of carbon atoms from alkyl chain (n_C) is described as follows:

$$\log k = \alpha n_C + \beta \quad (1)$$

where the parameters α and β are dependent on the nature of the anion part of the ionic liquids [50]. The dependence of the retention factor of a certain ionic liquid on the mobile phase composition is described as follows:

$$\log k = a + b\,C_{\text{o.m.}} \quad (2)$$

where $C_{\text{o.m.}}$ represents the concentration of the organic modifier from the mobile phase (as percentage), and a and b are regression parameters calculated from the linear representation between the logarithm of the experimental retention factor and different mobile phase compositions used for the retention study. According to Eq. (2), the retention of IL species can be enhanced by decreasing the organic modifier (acetonitrile, methanol, and *i*-propanol) content in the mobile phase.

Another possibility to separate ionic liquids is the utilization of the ion-pairing mechanism. Owing to their positive charge, these ionic species can form ion pairs with opposite charged species, such as, for example, the alkylsulphonate anions, which can be interact by van der Waals forces with hydrophobic stationary phases (C8, C18, or even pentafluorophenylpropyl silica-based stationary phase) [51].

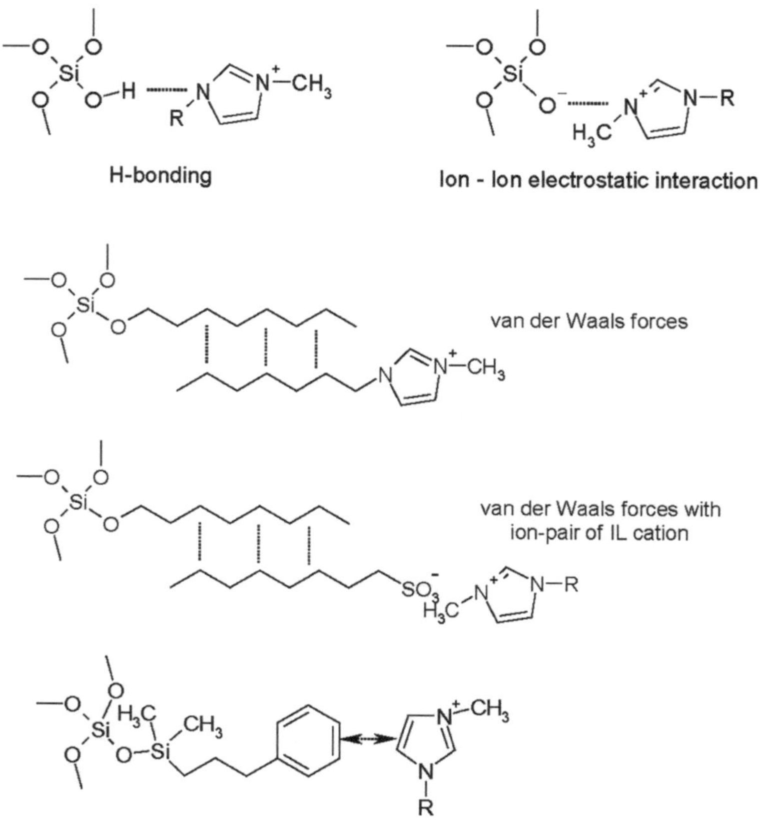

Figure 12.
Common intermolecular interactions between IL cation and silica gel (electrostatic interactions), C18 silica gel (van der Waals forces), and phenyl silica gel stationary phases (by cation-π interactions).

There are two models explaining the retention of ILs on hydrophobic stationary phases by the ion-pairing mechanism. The partition model supposes the formation of ion pair between IL+ and alkylsulphonate anion in the bulk of mobile phase, followed by the partition of the ion pair between the two phases, as represented by the following equilibria:

$IL^+ + CH_3\text{-}(CH_2)_n\text{-}SO_3^- \leftrightarrows IL^{+-}O_3S\text{-}(CH_2)_n\text{-}CH_3$.

$(IL^{+-}O_3S\text{-}(CH_2)_n\text{-}CH_3)_{mo.ph.} \leftrightarrows (IL^{+-}O_3S\text{-}(CH_2)_n\text{-}CH_3)_{st.ph.}$.

The adsorption model [52] considers that the alkylsulphonate anion is adsorbed on the stationary phase surface due to its hydrophobic chain, and the resulted charged stationary phase can interact electrostatically with the IL cation, according to the following simple equilibria:

$[CH_3\text{-}(CH_2)_n\text{-}SO_3^-]_{mo.ph.} \leftrightarrows [CH_3\text{-}(CH_2)_n\text{-}SO_3^-]_{st.ph.}$

$(IL^+)_{mo.ph.} + [CH_3\text{-}(CH_2)_n\text{-}SO_3^-]_{st.ph.} \leftrightarrows [CH_3\text{-}(CH_2)_n\text{-}SO_3^-IL^+]_{st.ph.}$.

In many cases, the graphical representations between the base 10 logarithm of the retention factor (log k) for the ionic liquid and the volume percentage of methanol in mobile phase ($C_{o.m.}$) showed a characteristic "U" shape with a minimum value within the studied interval of mobile phase composition. This experimental shape can be mathematically described by a second-order polynomial equation between log k and $C_{o.m.}$, as follows:

$$\log k = a + bC_{o.m.} + cC_{o.m.}^2 \tag{3}$$

where a is the intercept representing the extrapolated value of the retention factor (k_w), which corresponds to a mobile phase composed of only the aqueous component (a = log k_w for $C_{o.m.}$ = 0), while b and c can be calculated from the polynomial regression applied to this functional dependence. Mathematically, this dependence has a minimum point characterized by the value of $C_{o.m.}$, denoted by C^{min}, and the value of the retention factor, that is, min(log k), according to following equations:

$$C^{min} = -\frac{b}{2c} \text{ and } \min(\log k) = a - \frac{b^2}{4c} \tag{4}$$

This minimum corresponds to the mobile phase composition that allows the lowest affinity of the ion pairs between studied ionic liquids and alkylsulphonate for the octadecyl silica surface. Some experimental curves for the retention behavior of two ILs by ion pairing with C6, C7, and C8 alkylsulphonate anions [53] are illustrated in **Figure 13**.

Ion chromatography is another HPLC technique used for the analysis of ionic liquids [54]. Depending on which part of ILs is analyzed, the selection of the stationary phase is done to separate anions or cations. For example, the mixtures of ionic liquids with tetrafluoroborate $[BF_4]^-$, hexafluorophosphate $[PF_6]^-$, and bis(trifluoromethylsulfonyl) imide (triflimide) $[NTf_2]^-$ anions combined with several cations based on imidazole, pyridine, and tetrahydrothiophene could be analyzed for the anion purity without any influence of IL cation, by means of anion chromatography, using as eluent Na_2CO_3/$NaHCO_3$ in water-acetonitrile solutions [55]. The separation of IL cation can be performed by stationary phases with strong cationic exchanging properties, but the interaction between IL cations and stationary phase is rather complex, depending on the composition of mobile phase. The elution of mixtures of IL cations at different mobile phase compositions revealed a separation mechanism based on cation exchange, combined with nonspecific hydrophobic interactions [56].

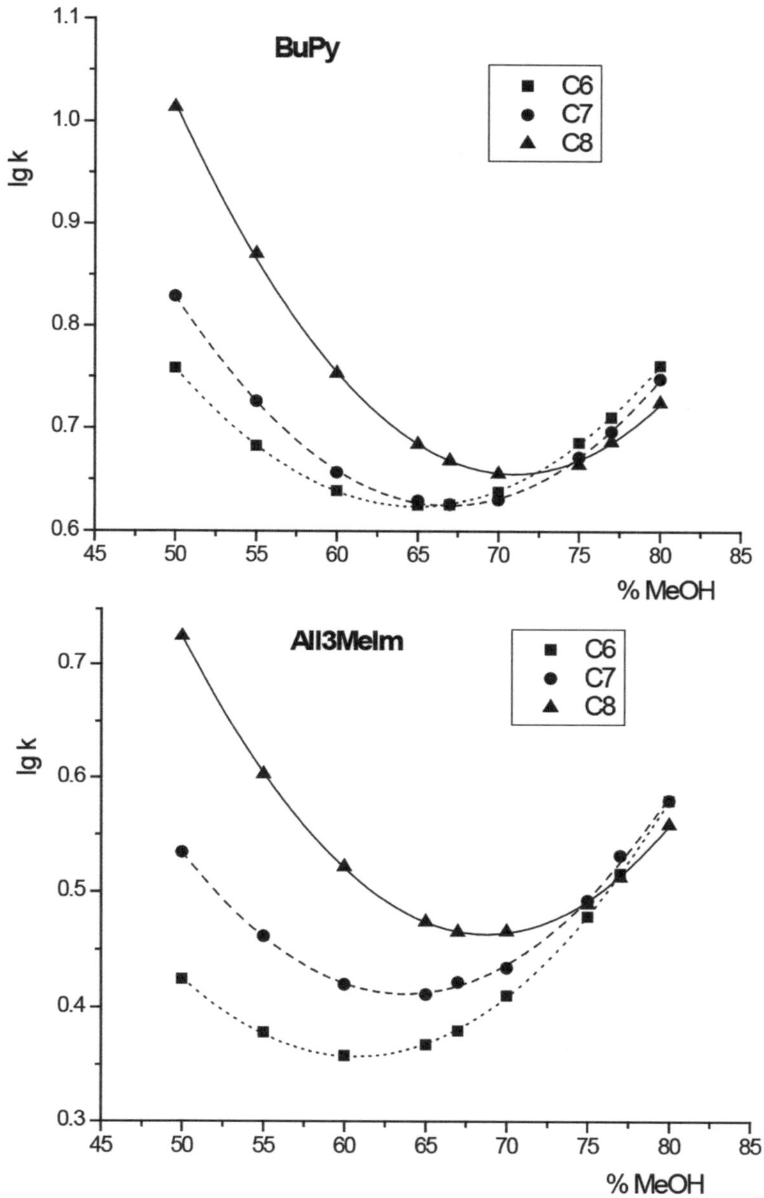

Figure 13.
Comparative dependences for log k versus $C_{o.m.}$, for the three alkylsulphonates as ion pairing agents, depicted for 1-butylpiridynium ($BuPy^+$) bromide and 1-allyl-3-methylimidazolium ($All3MeIm^+$) chloride [53].

5. Perspectives in utilization of ILs as materials for HPLC

Ionic liquids are more and more of interest in separation science. Analytical methods in high-performance liquid chromatography make use of these compounds as stationary phases having them on their surface as distinct functionalities For this purpose, more innovative stationary phases are synthetized by various approaches, involved in almost all the separation mechanisms in HPLC. Owing to the more complex structure of the IL-based stationary phases, the separation mechanism is more complex than with conventional stationary phases for reversed-phase, hydrophilic, chiral, or ionic HPLC. Ionic liquids are known as additive components in the mobile phase, and in many situations, they have improvements on the separation performances by the reduction of silanophilic interactions in liquid chromatography [57]. Owing to their charge, these additives may play the role of ion-pairing reagent and enhance the retention of dissociable compounds in reversed-phase HPLC. The interest in this class of compounds can be seen from the more increasing number of publications (research articles, reviews, and book chapters) in the last decade. This is why their perspective in liquid chromatography is not only a hope [58] but also a necessity for solving different complex problems with already proved advantages. Combinations with other materials may lead to more promising materials useful in separation applications [59].

Author details

Victor David[1*] and Serban C. Moldoveanu[2]

1 Department of Analytical Chemistry, University of Bucharest, Faculty of Chemistry, Bucharest, Romania

2 R.J. Reynolds Tobacco Co., Winston-Salem NC, USA

*Address all correspondence to: victor.david@chimie.unibuc.ro

References

[1] Ho TD, Zhang C, Hantao LW, Anderson JL. Ionic liquids in analytical chemistry: Fundamentals, advances, and perspectives. Analytical Chemistry. 2014;**86**:262-285

[2] McDaniel JG, Yethiraj A. Understanding the properties of ionic liquids: Electrostatics, structure factors, and their sum rules. Journal of Physical Chemistry B. 2019;**16**:3499-3512

[3] Zhang S, Sun N, He X, Lu X, Zhang X. Physical properties of ionic liquids: Database and evaluation. Journal of Physical Chemistry Reference Data. 2006;**35**(4):1475-1517

[4] Toh SLI, McFarlane J, Tsouris C, DePaoli DW, Luo H, Dai S. Room-temperature ionic liquids in liquid-liquid extraction: Effects of solubility in aqueous solutions on surface properties. Solvent Extraction and Ion Exchange. 2006;**24**:33-56

[5] Ventura SPM, Silva FA, Quental MV, Mondal D, Freire MG, Coutinho JAP. Ionic-liquid-mediated extraction and separation processes for bioactive compounds: Past, present, and future trends. Chemical Reviews. 2017;**117**(10): 6984-7052

[6] Sun P, Armstrong DW. Review: Ionic liquids in analytical chemistry. Analytica Chimica Acta. 2010;**661**(1):1-16

[7] Nawala J, Dawidziuk B, Dziedzic D, Gordon D, Popiel S. Applications of ionic liquids in analytical chemistry with a particular emphasis on their use in solid-phase microextraction. TrAC - Trends in Analytical Chemistry. 2018;**105**:18-36

[8] Nan H, Anderson JL. Ionic liquid stationary phases for multidimensional gas chromatography. TrAC - Trends in Analytical Chemistry. 2018;**105**:367-379

[9] Poole CF, Poole SK. Ionic liquid stationary phases for gas chromatography. Journal of Separation Science. 2011;**34**(8): 888-900

[10] Zhang Q. Ionic liquids in capillary electrophoresis for enantioseparation. TrAC - Trends in Analytical Chemistry. 2018;**100**:145-154

[11] Holzgrabe U, Wahl J. Ionic liquids in capillary electrophoresis. Methods in Molecular Biology. 2016;**1483**:131-153

[12] Feng J, Loussala HM, Han S, Ji X, Li C, Sun M. Recent advances of ionic liquids in sample preparation. TrAC - Trends in Analytical Chemistry. 2020;**125**:115833

[13] Koishi T. Molecular dynamics study of the effect of water on hydrophilic and hydrophobic ionic liquids. Journal of Physical Chemistry B. 2018;**122**(51): 12342-12350

[14] Rodrigues RF, Freitas AA, Canongia Lopes JN, Shimizu K. Ionic liquids and water: Hydrophobicity vs hydrophilicity. Molecules. 2021;**26**:7159

[15] Silva W, Zanatta M, Ferreira AS, Corvo MC, Cabrita EJ. Revisiting ionic liquid structure-property relationship: A critical analysis. International Journal of Molecular Sciences. 2020;**21**:7745

[16] Shi X, Qiao L, Xu G. Recent development of ionic liquid stationary phases for liquid chromatography. Journal of Chromatography A. 2015;**1420**:1-15

[17] Auler LMLA, Silva CR, Collins KE, Collins CH. New stationary phase for anion-exchange chromatography. Journal of Chromatography A. 2005;**1073**:147-153

[18] Wan X, Tian M, Row KH. Ionic liquid-modified silica as a new stationary phase for chromatographic separation. Journal of Analytical Chemistry. 2010;**65**: 816-820

[19] Van Meter DS, Oliver NJ, Carle AB, Dehm S, Ridgway TH, Stalcup AM. Characterization of surface-confined ionic liquid stationary phases: Impact of cation and anion identity on retention. Analytical and Bioanalytical Chemistry. 2009;**393**:283-294

[20] Wang Q, Baker GA, Baker SN, Colon LA. Surface confined ionic liquid as a stationary phase for HPLC. The Analyst. 2006;**131**:1000-1005

[21] Pino V, Alfonso AM. Surface-bonded ionic liquid stationary phases in high-performance liquid chromatography - a review. Analytica Chimica Acta. 2012;**714**:20-37

[22] Qiao L, Dou A, Shi X, Li H, Shan Y, Lu X, et al. Development and evaluation of new imidazolium-based zwitterionic stationary phases for hydrophilic interaction chromatography. Journal of Chromatography A. 2013;**1286**:137-145

[23] Huang Y, Yao S, Song H. Application of ionic liquids in liquid chromatography and electrodriven separation. Journal of Chromatographic Science. 2013;**51**:739-752

[24] Moldoveanu SC, David V. Essentials in Modern HPLC. Amsterdam: Elsevier; 2013

[25] Xin B, Hao J. Imidazolium-based ionic liquids grafted on solid surfaces. Chemical Society Reviews. 2014;**43**:7171-7187

[26] Svec F, Fréchet JMJ. Continuous rods of macroporous polymer as high-performance liquid chromatography separation media. Analytical Chemistry. 1992;**64**:820-822

[27] Wang QC, Svec F, Fréchet JMJ. Macroporous polymeric stationary phase rod as continuous medium for reversed-phase chromatography. Analytical Chemistry. 1993;**65**:2243-2248

[28] Steinke JHG, Dunkin IR, Sherrington DC. Transparent macroporous polymer monoliths. Macromolecules. 1996;**29**:5826-5834

[29] Zhu T, Bi W, Row KH. A new ionic liquids-based monolithic column for determination of caffeine and theophylline. Journal of Applied Polymer Science. 2010;**118**(6):3425-3430

[30] Qiu H, Mallik AK, Takafuji M, Jiang S, Ihara H. New poly(ionic liquid)-grafted silica multi-mode stationary phase for anion-exchange/reversed-phase/hydrophilic interaction liquid chromatography. The Analyst. 2012;**137**: 2553-2555

[31] Qiu H, Takafuji M, Sawada T, Liu X, Jiang S, Ihara H. New strategy for drastic enhancement of selectivity *via* chemical modification of counter anions in ionic liquid polymer phase. Chemical Communications. 2010;**46**:8740-8742

[32] Qiu H, Zhang M, Chen J, Gu T, Takafuji M, Ihara H. Anionic and cationic copolymerized ionic liquid-grafted silica as a multifunctional stationary phase for reversed-phase chromatography. Analytical Methods. 2014;**6**:469-475

[33] Zhang M, Chen J, Qiu H, Mallik AK, Takafuji M, Ihara H. Homogenous formation and quaternization of urea-functionalized imidazolyl silane and its immobilization on silica for surface-confined ionic liquid stationary phases. RSC Advances. 2014;**4**:34654-34658

[34] Qiao L, Shi X, Lu X, Xu X. Preparation and evaluation of surface-bonded tricationic ionic liquid silica as

stationary phases for high-performance liquid chromatography. Journal of Chromatography A. 2015;**1396**:62-71

[35] Liu H, Guo Y, Wang X, Liang X, Liu X. Amino-terminated ionic liquid modified grapheme oxide coated silica composite stationary phase for hydrophilic interaction chromatography. RSC Advances. 2014;**4**:37381-37388

[36] Liang X, Chen Q, Liu Q, Jiang S. Comparison of liquid chromato-graphic behaviors on methylimidazolium functionalized ZrO2/SiO2-4 and N-methylimidazolium functionalized SiO2 stationary phases. Journal of Chromatography A. 2008;**1182**:197-204

[37] Zhou Z, Li X, Chen X, Hao X. Synthesis of ionic liquids functionalized β-cyclodextrin-bonded chiral stationary phases and their applications in high-performance liquid chromatography. Analytica Chimica Acta. 2010;**678**:208-214

[38] Liu D, Wang H, Liang M, Nie Y, Liu Y, Yin M, et al. Polymerized phosphonium ionic liquid functionalized silica microspheres as mixed-mode stationary phase for liquid chromatographic separation of phospholipids. Journal of Chromatography A. 2021;**1660**:462676

[39] Moldoveanu SC, David V. Selection of the HPLC Method in Chemical Analysis. Amsterdam: Elsevier; 2017

[40] Marszal MP, Baczek T, Kaliszan R. Reduction of silanophilic interactions in liquid chromatography with the use of ionic liquids. Analytica Chimica Acta. 2005;**547**:172-178

[41] Zhang W-Z, He L-J, Gu Y, Liu X, Jiang S-X. Effect of ionic liquids as mobile phase additives on retention of catecholoamines in reversed-phase high-performance liquid chromatography. Analytical Letters. 2003;**36**:827-838

[42] Han D, Wang Y, Jin Y, Row KH. Analysis of some β-lactam antibiotics using ionic liquids as mobile phase additives by RP-HPLC. Journal of Chromatographic Science. 2011;**49**:63-66

[43] Zhang W-Z, He L-J, Liu X, Jiang S-X. Ionic liquids as mobile phase additives for separation of nucleotides in high-performance liquid chromatography. Chinese Journal of Chemistry. 2004;**22**: 549-552

[44] He L-J, Zhang W-Z, Zhao L, Liu X, Jiang S-X. Effect of 1-alkyl-3-methylimidazolium-based ionic liquids as the eluent on the separation of ephedrines by liquid chromatography. Journal of Chromatography A. 2003;**1007**:39-45

[45] Antonio VH, Javier H, Miguel AR. Ionic liquids as mobile phase additives for the high-performance liquid chromatographic analysis of fluoroquinolone antibiotics in water samples. Analytical and Bioanalytical Chemistry. 2008;**392**:1439-1446

[46] Treder N, Oledzka I, Roszkowska A, Baczek T, Plenis A. Control of retention mechanisms on an octadecyl-bonded silica column using ionic liquid based mobile phase in analysis of cytostatic drugs by liquid chromatography. Journal of Chromatography A. 2021;**1651**:462257

[47] Galaon T, Radulescu M, David V, Medvedovici A. Use of an immiscible diluent in ionic liquid/ion pair LC for the assay of an injectable analgesic. Central European Journal of Chemistry. 2012;**10**(4):1360-1368

[48] Jiang Q, Qiu H, Wang X, Liu X, Zhang S. Effect of ionic liquids as additives on the separation of bases and amino acids in HPLC. Journal of Liquid Chromatography & Related Technologies. 2008;**31**(10):1448-1457

[49] Studzinska S, Buszewski B. Study of retention mechanism of imidazolium-based ionic liquids in HPLC. Journal of Separation Science. 2010;**33**(9):1264-1273

[50] Ruiz-Angel MJ, Berthod A. Reversed phase liquid chromatography of alkyl-imidazolium ionic liquids. Journal of Chromatography A. 2006;**1113**:101-108

[51] Nichthauser J, Stepnowski P. Retention mechanism of selected ionic liquids on a pentafluorophenylpropyl polar phase: Investigation using RP-HPLC. Journal of Chromatographic Science. 2009;**47**:247-253

[52] Bartha A, Stahlberg J. Electrostatic retention model of reversed-phase ion-pair chromatography. Journal of Chromatography. A. 1994;**668**:255-284

[53] Radulescu M, Iorgulescu EE, Mihailciuc C, David V. Comparative study of the adsorption of pyridinium and imidazolium based ionic liquids on octadecylsilica stationary phase under ion pairing mechanism with alkylsulphonate anions. Revue Roumaine de Chimie. 2012;**56**(1):61-67

[54] Paszkiewicz M, Stepnowski P. How should ionic liquids be analyzed. Current Organic Chemistry. 2011;**15**(12):1873-1887

[55] Rutz C, Schmolke L, Gvilava V, Janiak C. Anion analysis of ionic liquids and ionic liquid purity assessment by ion chromatography. Journal of Inorganic and General Chemistry. 2017;**643**(1):130-135

[56] Stepnowski P, Mrozik W. Analysis of selected ionic liquid cations by ion exchange chromatography and reversed-phase high performance liquid chromatography. Journal of Separation Science. 2005;**28**(2):149-152

[57] Marszal MP, Kaliszan R. Application of ionic liquids in liquid chromatography. Critical Reviews in Analytical Chemistry. 2007;**37**:127-140

[58] Ali I, Suhail M, Sanagi MM, Aboul-Enein HY. Ionic liquids in HPLC and CE: A hope for future. Critical Reviews in Analytical Chemistry. 2017;**47**(4):332-339

[59] Hui BY, Raoov M, Zain NDM, Mohamad S, Osman H. Combination of cyclodextrin and ionic liquid in analytical chemistry: Current and future perspectives. Critical Reviews in Analytical Chemistry. 2017;**47**(5):454-467

Chapter 5

Perspective Chapter: Advantages of Ion Mobility Coupled with HPLC/UPLC

Robert Owen Bussey III

Abstract

Ion mobility is a new separation technique that can be coupled with high performance liquid chromatography (HPLC) or ultra-performance liquid chromatography (UPLC). Variances in cross-sectional ionic areas of different molecules create differential speeds through a gas allowing for millisecond separations. Combining ion mobility with both liquid chromatography and mass spectrometry with fragmentation, separations can be achieved on the second (HPLC), millisecond (ion mobility), and microsecond (mass spectrometry) timescales. This orthogonal separation greatly cleans up mass spectral data of co-eluting peaks from the liquid chromatography and adds to the descriptive data of each ion. With descriptive data such as retention time, cross-sectional area, m/z ratio, and mass spectral fragmentation, many options become available for analytical analysis. Options ranging from descriptive data collation into instrument libraries to sensitivity enhancement for trace analysis will be explored in this chapter along with the description of different forms of ion mobility.

Keywords: ion mobility, UPLC, HPLC, liquid chromatography, mass spectrometry, fragmentation, collision cross section, CCS

1. Introduction

This chapter will explain the basic principles around ion mobility along with some different forms of ion mobility and how they function. Advantages and disadvantages of each technique will be offered. Case studies will demonstrate the effectiveness of using ultra-performance liquid chromatography (UPLC) paired with ion mobility and mass spectrometry. The drift time or the time required to traverse the ion mobility cell can be used to align and differentiate the mass spectra of isobaric species in complex matrices. This differentiation may allow for better quantitation and/or sensitivity enhancement of those features. In addition, assignment of descriptive data such as collision cross sections can be used to form databases for targeted and untargeted analysis.

2. Ion mobility explained

One of the most common applications of ion mobility in everyday life is swabbing luggage and hands at airports for nitrate-based explosives [1]. It is portable and fast with high volume and highly reproducible results. The fundamental concept of ion mobility is ionic separation in a gas with an applied electric field. As the ions are propelled down a drift tube by an electrical field, the ions hit the gas molecules, and their velocity slows down based on the number of gas molecules that interact with the ions [2]. **Figure 1** shows the basic concept behind the ability of different sized ions moving at different rates through a drift tube. Drift time or the time it takes ions to travel the entire length of the drift tube would be the observed measurement from the ion mobility spectrometer. The drift time and other parameters can be used to calculate the collision cross section (CCS) of each compound to be used as compound-specific descriptive data. The larger the cross-sectional area, the more interaction there are with the gas molecules (lower mobility) and the slower the ion moves through the drift tube. When the cross-sectional area of the ion is smaller, there is less interaction with the gas molecules (higher mobility), and the ion can move faster compared with the larger ions [2, 3]. The next sections will describe different types of ion mobility spectrometers along with reported advantages and disadvantages of each technique. The details of each technique allow for a more informed decision when purchasing a system with a specific ion mobility separation technique. Important facts to think about would be isomer separation, ion mobility aligned spectra, and CCS fingerprinting for databases [2].

3. Types of ion mobility

3.1 DTIMS

Drift tube ion mobility spectrometry (DTIMS) is potentially the simplest of the ion mobility techniques based on a simplified structural and parameter interplay [2, 4]. The

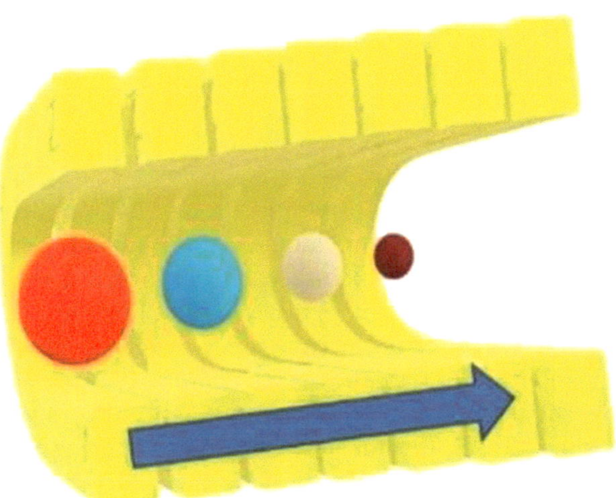

Figure 1.
Ion mobility chamber separating ions based on mobility.

drift tube is filled with a defined volume of buffer gas with no directional flow, which acts as a velocity regulator to moving ions [4]. The ions are propelled through the drift tube by a uniformly applied, static electric field with a decreasing voltage as the ions traverse the drift tube [2, 4]. As seen in **Figure 1**, the drift tube is made up of stacked electrodes that allow for the electric field to be applied over the length of the tube. The static electric field, defined drift tube length, and time it takes to traverse the entire length of the drift tube allow for ionic separation and CCS calculations [4]. One of the disadvantages of the DTIMS system is the voltage drop across the length of the tube. In order to create a good separation, the tube length can be increased, but that would require a higher electric field at the beginning of the drift tube to propel the ions a greater distance [2, 5].

There are many advantages to DTIMS. Unlike traveling wave ion mobility spectrometry (TWIMS), DTIMS does not require calibration with a complex mixture of compounds with well-defined CCS values in order to reproducibly measure CCS values [6]. The well-defined CCS values for TWIMS calibration were first acquired on a DTIMS instrument. In some instances, CCS calibration is required, but no complex mixture is needed to increase CCS reproducibility in DTIMS. Even without calibration, a comprehensive drift time library of many analytes can be collected in one experiment [2].

This comprehensive drift time library is limited by the duty cycle of the DTIMS systems. The duty cycle or the time in between ion trapping, ion separation, ion detection, and another cycle is shorter compared with ion mobility techniques that do not trap [6]. This trapping and analysis of ion groups limit how many ions that can be separated at any one time versus a continuous stream of ions entering the ion mobility system [7]. Instead of waiting for the duty cycle to finish, some vendors have been sending packets of ions into the drift tube one after the other before the end of the duty cycle in order to measure overlapping drift time experiments [8].

3.2 TWIMS

The traveling wave ion mobility spectrometry (TWIMS) drift tube is structurally similar to the DTIMS tube with stack electrodes propelling an ion with an electric field through a fixed volume of buffer gas [2, 6]. The electrical field oscillates continuously with no voltage drop versus a static electrical field with decreasing voltage over the length of the tube. This oscillation creates oscillating voltage waves that push the ion through the drift tube in a manner similar to surfers on the top of waves in the ocean [6]. The top of the wave carries the ion for a period until the mobility of the ion slows its velocity and allows the wave to pass it. The ion continues to be propelled by subsequent waves until it reaches the detector. An ion with a small collision cross section will ride the wave farther than ions with larger collision cross sections. This also means that ions with larger collision cross sections will require more waves to push it through the drift tube due to the slower velocity [6].

Unlike the DTIMS, the TWIMS needs to be calibrated with a complex mixture of ions that have known drift times. This allows for a continuous measurement of the ions with predetermined parameters based on the calibrated standards. Most likely the standards would include compounds with a wide range of CCS values along with a wide mass range [6, 9]. The ion mobility pressure chamber is locked before calibration, and if pressure changes occur after calibration, then the calibration will have to be repeated to account for the slight pressure change [9]. If there are slight differences in CCS values between instruments, the fundamental parameters of ion mobility should be compared.

Some important advantages of TWIMS are low-voltage requirements due to a constant wave height in an oscillating electrical field and voltage modulation over long drift tube lengths to maintain ion flow [2, 10]. Both of these qualities allow for the movement of ions across longer distances, which increase the interactions with the buffer gas and increase peak resolution. Longer path lengths would not be possible without the low-voltage requirements. Remember that the DTIMS has a static electric field with a linearly decreasing voltage over the length of the tube. If a DTIMS tube length was increased, the voltage would also increase in order provide enough momentum to push the ions through the drift tube into the detector [11].

Cyclic ion mobility systems are an extension of the TWIMS system except the length of the drift tube has the potential to vary based on how many circular passes it takes in the cyclic mobility tube. The hardware has ion guides directing ions to the circular ion mobility cell perpendicular to the main body of the system. It is similar to a trap and allows the user to choose how many circular revolutions it will take. This ability to customize the drift tube length allows for better separation/customization and has the potential to create ion mobility separation far surpassing previous instrument resolution [12].

3.3 TIMS

Trapped ion mobility spectrometry (TIMS) is a relatively new commercialized product. It uses a nonuniform electrical field unlike DTIMS to trap ions. Both DTIMS and TWIMS have constant gas volumes with no flow with the only movement caused by the electric field and voltage changes. TIMS uses both buffer gas flow toward the detector and electric field changes over the length of the ion mobility cell to propel the ions to the MS detector [13, 14]. The electric field can be tuned to guide the ions to the detector.

TIMS is composed of three regions: entrance funnel, ion mobility tube, and the exit funnel. The entrance funnel focuses and compacts the ions using an ion funnel into the ion mobility cell [15]. When the ions enter the mobility cell after focusing, the DC electric field at the exit is set higher than the potential in the mobility cell and at a 180-degree angle. This creates a field, which repeals the ions near the exit. The ions are trapped because the air flow pushes them toward the exit, and the DC field pushes them away from the exit [15]. In addition, a low-intensity electric field starts to gradually increase over the length of the mobility cell. The ions are trapped and separated based on their size-to-charge ratio [13, 15]. The ions with the lowest mobility or the largest size-to-charge ratio will congregate toward the exit with the highest electric field. This is caused by the gas molecules interacting more with the largest compounds, thus pushing them closer to the exit. The compounds with the highest mobility or the smallest size-to-charge ratio will be farther from the exit because of the opposing field at the exit pushing it away [13]. The smaller ions interact less with the gas molecules causing them to move less downstream and more away from the exit. The ions can exit the mobility cell by lowering the electric field intensity gradually at the exit. This will cause the lowest mobility ions to exit first due to the gas flow and the highest-mobility compounds last. The gradual decrease in the electric field intensity allows for a segmented elution of compounds based on the parameters of the experiment. Having the lower-mobility compounds come out first is opposite to the behavior of DTIMS and TWIMS [15].

Unlike DTIMS and TWIMS, which allow for scanning to see multiple ion mobilities with the same experimental conditions, TIMS requires sequential experimental parameter changes to see multiple ions injected into the mobility chamber. As stated above, there is a gradual decrease in the electric field at the exit, which causes a segmented release of the ions [13]. This change in voltage only allows certain mobilities to exit at a time without allowing higher-mobility ions to exit at the same voltage. TIMS is a highly selective technique relating to separation efficiency, but the ability to scan is lost or needs extra parameter changes. The trapping ability allows the TIMS to maintain peak separation despite short path lengths and short lab instrument footprints [15].

3.4 FAIMS

Field asymmetric ion mobility spectrometry (FAIMS) is an atmospheric pressure ion mobility technique. This technique uses both high and low electric fields to separate ions in the gas-phase mobility cell. The FAIMS device is small, and it can potentially be placed in between the ion source and the vacuum inlet of the mass spectrometer [2, 16]. The integration onto MS systems without ion mobility is achieved by small accessories to maintain the required negative pressure in the mass spectrometer. Like TIMS, gas flow is used for ion movement in addition to the electric field. In FAIMS, the electric field alternates both in strength and polarity to separate ion according to field strength changes and not drift time mobilities over the length of the mobility cell [16].

FAIMS cannot provide CCS values due to a lack of mobility measurement and change in ion structure due to the changes in electric field strength and polarity. This technique cannot scan for ions with multiple electric field strength changes but rather specific electric field changes. This can allow a continuous monitoring of ions with the same electric field change response. For compounds belonging to similar chemical classes, this would be beneficial because the signal-to-noise ratio would increase [2, 17].

4. Case studies

4.1 Case study 1

The Yassin et al. [18]study pairs ion mobility, UPLC and mass spectrometry together allowing for the characterization of many polyphenolic compounds in tea. The reversed-phase UPLC separation yielded chromatographic peaks, but when coupled with IMS separation, additional peaks were revealed. This section will summarize how ion mobility, liquid chromatography, and MS fragmentation can be used to characterize structural features of unknowns.

The data were acquired in negative mode using MS^e mode to collect low- and high-collision energy spectra during the same acquisition. The traveling wave ion mobility was used to collect drift times and calculate collision cross sections. The drift time, pseudo-molecular ion m/z, and mass spectral fragments were then aligned to the LC elution time. This is very useful in complex mixtures because mass fragments of two isobaric species may overlap with no indication to which pseudmolecular ion the fragments belong.

Theasinensin C (TS) and proanthocyanidin B (PA) were used as examples of polyphenolic isomeric species in this study (**Figure 2**). Assam and Sri Lanka teas were analyzed for the presence of TS isomers. The levels of PA were compared in extracts of *Ziziphus spina-christi* and *Rhododendron*. The extracted ion chromatograms of each extract used a nominal mass search of 609 Da with resulting accurate masses ranging from 609.156 Da to 609.183 Da. Characteristic fragmentation patterns of TS and PA standards were created for family classification of any potential new isomers.

The Assam tea extract benefited from having UPLC and ion mobility together. Ion mobility was able to show that a 609 Da feature seen at 1.28 min contained two compounds with distinct drift times of 5.18 ms and 5.56 ms and two distinct fragmentation patterns characteristic of the TS family. In this case, the ion mobility was able to separate the features the UPLC could not. In addition, the UPLC separation showed

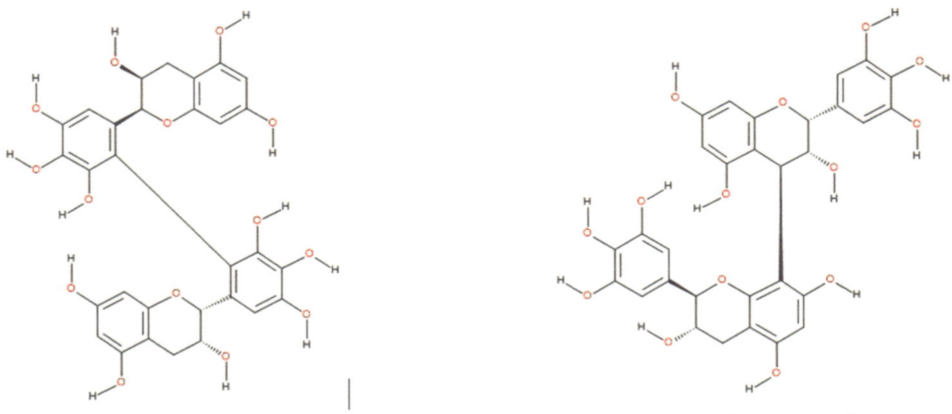

Theasinensin C | Proanthocyanidin

Figure 2.
Structural differences in isomers Theasinensin C and Proanthocyanidin.

additional 609 Da features at 10.09 min and 10.22 min with the same drift time of 5.67 min. Without the UPLC separation, these two compounds would have been classified as one isomer and not two. The same features were found in the Sri Lanka tea.

According to this investigation, drift times close in value and at the same LC retention time were investigated further. Under normal conditions, the gas is heated in the ion mobility tube, and this elevated temperature may cause structural changes between two isomeric forms. There would be an equilibrium between both diasterioisomers each with its own similar but distinct collision cross section, which would cause two peaks to appear in the ion mobility plot. The investigators looked at the MS^e fragmentation of the TS isomers at both 5.18 ms and 5.56 ms. The isomer at 5.56 ms had a loss of 18 Da characteristic of water, whereas the TS isomer at 5.18 ms did not lose water. PAs, which are similar in structure to TSs, have been shown to lose water from epicatechin moieties, whereas PAs with catechin moieties do not loose water. To test whether this was also true with the TS family, the investigators employed molecular modeling. A correlation was established between the isomer drift times and their calculated collisional cross sections. The data showed that the isomer with 5.18 ms drift time had a collisional cross section of 154 $Å^2$ and a trans stereochemistry from a catechin building block. The isomer at 5.56 ms had a calculated collisional cross section of 157 $Å^2$ with a cis stereochemistry from an epicatechin moiety. The confirmed epimerization was a drawback of ion mobility in this study because the TS could not definitively be assigned a cis or trans designation at retention time of 1.28 min.

The proanthocyanidins (PAs) in both the *Ziziphus spinae-christi* and *Rhododendron* extracts showed interesting results for UPLC separation with ion mobility. The *Ziziphus spinae-christi* extract showed two 609 Da features at 1.67 min and 10.22 min with specific drift times at 5.35 ms and 5.62 ms. Fragmentation confirmed PA isomers. The similar drift times would have required further investigation like with the example above, but since the peaks were chromatographically well resolved, the peaks were not artifacts of the ion mobility. The *Rhododendron* extract showed 609 Da features at 2.35 min and 4.98 min with specific drift times of 5.24 ms and 8.37 ms. Fragmentation confirmed very similar PA isomeric structures with different drift times, and this suggested the presence of regioisomers. Putative identifications of regioisomers epigallocatechin-(4,8)-epigallocatechin (PA) and epigallocatechin- (4,6)-epigallocatechin were suggested. Computational modeling data agreed with the experimentally observed collisional cross sections of 154 Å^2 (5.24 ms drift time) for the epigallocatechin-(4,8)-epigallocatechin and 178 Å^2 (8.37 ms drift time) for the epigallocatechin-(4,6)-epigallocatechin (**Figure 3**).

The combination of ion mobility, UPLC chromatography, and mass spectral fragmentation allowed the separation of isomers that may not have been separated in any single technique. Excess heat in the ion mobility drift tube was shown to be one drawback to the technique because it caused epimerization. In addition, descriptive data provided by each technique helped guide structural confirmation.

4.2 Case study 2

A further look into pairing liquid chromatography, ion mobility, and mass spectrometry together for chemical profiling is described below in a work by McCullagh et al. [19]. Analytical approaches that provide the maximum amount of descriptive are needed for complex herbal extracts for product authentication, extract profiling, stability/degradation, and purity analysis. This study investigated the genus *Passiflora* for the many flavanoid derivatives such as C-glycosylflavone specifically apigenin/luteolin derivatives, which may have medicinal properties. Main areas of emphasis will be the analysis of 6-C and 8-C glycosides to establish exact mass fragmentation, LC chromatographic separation, and ion mobility collision cross sections (CCS) to decrease sample complexity in a herbal extract matrix. An additional goal was to establish a CCS searchable database with attached fragmentation and LC data. Fragmentation and LC data are more variable between systems, whereas the CCS values should be more reproducible.

The *Passiflora* species evaluated include *P. edulis*, *P. alata*, *P. incarnata*, and *P. caerula*. Each sample was extracted, and the flavone fraction was purified with solid-phase extraction. Reversed-phase chromatography was performed using C18 chromatography. The data were acquired in both positive and negative modes using MS^e to collect low- and high-collision energy spectra during the same acquisition. The low energy was 4 eV and the high energy was a ramp from 30 to 75 eV. Traveling wave ion mobility was used to collect drift times and calculate collision cross sections. The pseudomolecular ion m/z and mass spectral fragments were then aligned to the retention time and drift time to create a database with calculated CCS values.

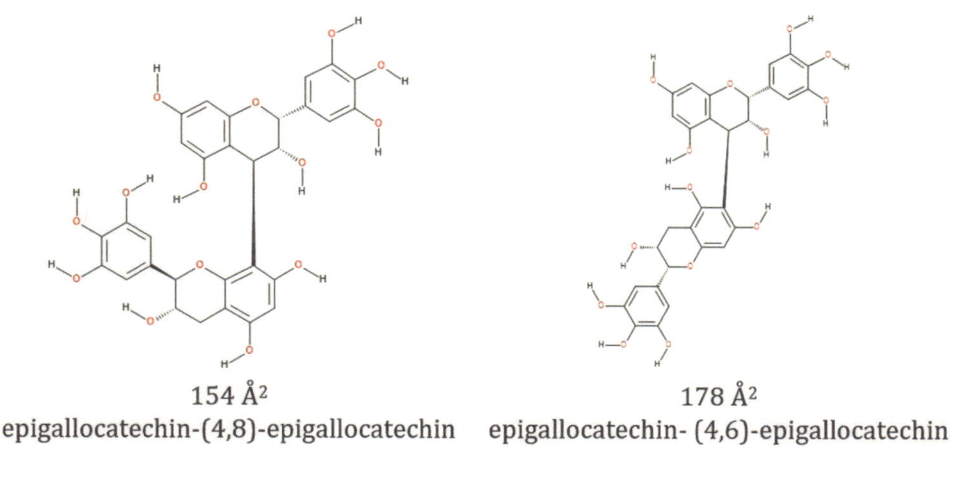

Figure 3.
Connectivity of proanthocyanidin isomers and different CCS values.

This study incorporated targeted profiling of flavanoid derivatives in the genus *Passiflora* to establish a non-targeted approach. The 6-C and 8-C-glycosylflavone isomers of orientin/isoorientin and vitexin/isovitexin had similar isomeric structures (**Figure 4**). Each standard was characterized both in pure solvent and plant extract matrix. All CCS values were based on data from replicate injections of pure standards at high and low concentrations to simulate different amounts in extract and different ionization responses in plant matrix. The CCS values created in the study were used as metrics to confirm the presence of isomeric flavanoids similar to the standards. The identity of these unknown flavanoids would not need to be known since the descriptive data were assigned to each unknown during this study.

The customized database created in this study aligned accurate mass of the pseudomoleulcar ions and fragment ions to retention time and drift time. This drift time alignment helped assign fragments to the correct pseudomolecular ion. Collision cross section values were created to use with a delta $^{TW}CCSN_2$ metric to help to differentiate flavanoids despite co-elution. Based on this study, this delta $^{TW}CCSN_2$ metric can be used even on trace components when fragmentation data would be hard to attain.

The CCS values and retention time of each isomeric pair orientin/isoorientin and vitexin/isovitexin are displayed in **Table 1**. Notice that the retention times of the isomers within each pair are very close, and without drift time alignment, the mass spectral fragmentation would be blended without specific assignment to either isomer. These values were then used to calculate delta $^{TW}CCSN_2$ taking the difference between the CCS values of each isomer within each pair. The compounds with glycosides attached at carbon-8 had smaller CCS values compared with the compounds with glycosides attached at carbon-6 (**Table 1**). These values are determined by their 3D structure

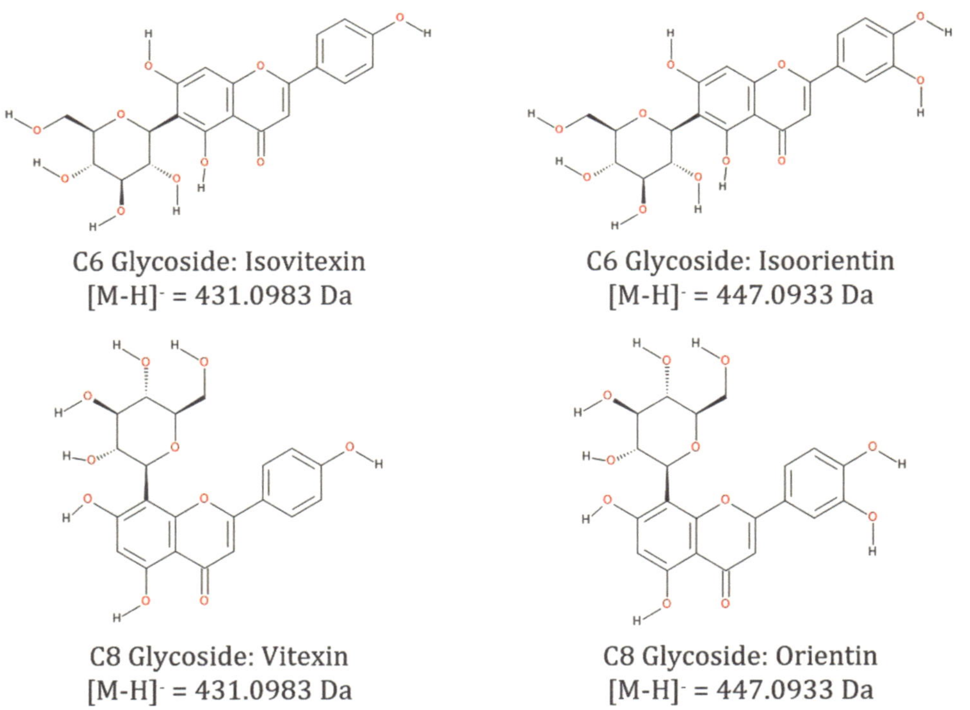

Figure 4.
Structure differentiation between C6 and C8 pairs isoorientin/orientin and isovitexin/vitexin.

either being more compact or spread out to change the interaction with the gas molecules as they travel through the drift tube. The CCS values for the negative mode were more different than those in the positive mode (**Table 1**). In the positive mode, the close chromatographic retention times and close CCS values would cause little to no baseline resolution in both dimensions. In this case, the negative mode data would allow for better differentiation of the isomers despite possible co-elution. In addition, the drift time aligned data would allow the correct quantitation of each isomer. The delta $^{TW}CCSN_2$ method was tested with *P. alata*, *P. incarnata*, and *P. caerulea*, and the isomers that were found were within 0.75% error. In addition, these results were replicated over 3 years, and the delta $^{TW}CCSN_2$ values were within 0.8% error for that time period. When this metric is used with accurate mass of parent and daughter ions and retention time, the database can be used to better analyze complex herbal extracts.

In the *Passiflora* matrix, the retention times of vitexin and isovitexin were based on the apexes of the peaks with no chromatographic baseline resolution. This was where the ion mobility aided in differentiating vitexin and isovitexin. With no baseline separation, the fragmentation patterns of both isomers were overlapping with no distinction between them. When the accurate masses of the parent and fragment ions were drift time aligned, this allowed the fragmentation patterns to be established for each of the isomers using the collision energy ramp of 30 eV–75 eV. The distinctive parts of the fragmentation were the ratios of m/z 281/282/282/284 to distinguish between vitexin and isovitexin. For vitexin, the characteristic ion ratios were m/z 281 < 282 and m/z 282 < 284. For isovitexin, the ion ratios were m/z 281 > 282 and m/z 282 > 284. The less than or greater than refers to the ion intensities. When these ions were retention time and drift time aligned, the identification of specific isomers was successful.

The retention times of orientin and isoorientin were within 0.05 min; therefore, there was no chromatographic baseline resolution. Again, the ion mobility aided in differentiating the isomers along with the elevated collision energy ramp of 30 eV–75 eV. The fragmentation patterns of both isomers were overlapping with no distinction between orientin and isoorientin. The accurate masses of the parent and fragment ions were drift time aligned allowing the fragmentation patterns to be established for each of the isomers. The distinctive parts of the fragmentation were the ratios of 284/285 and m/z 297/298/299 to distinguish between orientin and isoorientin. For orientin, the characteristic ion ratios were m/z 284 > 285, m/z 297 > 298, m/z 298 > 299. For isoorientin, the ion ratios were m/z 284 < 285, m/z 297 < 298, m/z 298 < 299. The fragments could be used as another metric to differentiate between orientin and isoorientin with retention time and drift time alignment. The increased specificity from CCS libraries may create better future characterization protocols for herbal extracts in consumer products.

Flavanoid	Orientin	Isoorientin	Vitexin	Isovitexin
Retention Time (min)	7.88	7.83	8.52	8.40
Negative Polarity CCS (A^2)	187.7	198.1	188.8	195.5
$\Delta^{TW}CCSN_2$ (A^2)	10.4		6.7	
Positive Polarity CCS (A^2)	200.7	203.4	198.2	199.3
$^{TW}CCSN_2$ (A^2)	2.7		1.1	

Table 1.
Retention times and CCS values for negative and positive MS experiments.

Combining accurate mass, fragmentation data, IMS separation, and CCS measurements created what the authors of this study called "known-unknown" fingerprinting. The normal protocol for herbal extract characterization is to test for a small number of pure active compounds, but often these standards are in limited supply or are cost prohibitive. Using this "known-unknown" fingerprinting technique, features can be cataloged. As more descriptive data are assigned to unknown features, putative identifications may be assigned. The unknowns can also be cataloged into the "known-unknown" database to be tracked from sample to sample even if there is no confirmed identification. This technique was used in the *P. caerulea* species with unknown analytes present with accurate masses ranging from 431.0958 Da to 431.0983 Da with CCS values ranging from 185.5 Å^2 to 188.0 Å^2. Collecting these data along with the fragmentation data added descriptive data to this type of non-targeted analysis for future comparison to other *Passiflora* species.

The "known-unknown" workflow was applied to compounds with the masses of 431.09 Da and 447.09 Da or compounds similar to the orientin/isoorietin and vitexin/isovitexin isomer pairs. Nineteen different candidates in *P. caerula*, *P. edulis*, *P. alata*, and *P. incarnata* were identified with similar accurate masses, fragmentation, and CCS values as the C6/C8 glycoside isomers. All the abovementioned descriptive data allowed putative identification of C6 or C8 glycosidic isomers. As an example, two isomers with similar retention times shared accurate mass of 431.098 Da, but one had a higher CCS value than the other. The higher CCS value would suggest a C6 glycoside, and to confirm, the fragmentation data were evaluated. The C6 glycoside had fragment ion ratios with $m/z\ 284 < 285$ and $m/z\ 297 < m/z\ 298 > m/z\ 299$. The C8 glycoside had fragment ratios with $m/z\ 284 > m/z\ 285$, and $m/z\ 297 > m/z\ 298 < m/z\ 299$. These unknowns were similar to isovitexin/vitexin, but not the same since the retention times were different than the standards.

Other than the "known-unknown" characterization, a system with ultra-performance liquid chromatography, ion mobility, and mass spectrometry (UPLC-IM-MS) can be used to better quantify convoluted peaks. When isobaric species are present, the peak areas overlap with no demarcation of either peak, but when ions are drift time aligned, isomeric quantitation can occur. The drift time alignment separated the peak areas of each isomer, allowing for the quantitation of each isomer even in the presence of a complicated matrix. Using this concept, the authors quantitated isoorientin, orientin, isovitexin, and vitexin in *P. caerula*, *P. edulis*, *P. alata*, and *P. incarnata*. To further show this concept, the authors calculated the concentrations with and without drift time alignment. As expected, the isobaric species had greatly different calculated concentrations with and without drift time alignment.

In this study, the authors demonstrated the effectiveness of using UPLC paired with ion mobility and mass spectrometry. The drift time alignments allowed for differentiation of mass spectra of isobaric species. This differentiation also allowed for better quantitation of isobaric species. The calculated CCS values aided in the formation of a database of "known-unknowns" that could be tracked between herbal extracts despite the unconfirmed identities of some compounds.

4.3 Case study 3

Adams et al. [20] created an liquid chromatography trapped ion mobility spectrometry with mass spectrometry (LC-TIMS-MS) technique to provide a high-throughput orthogonal separation technique for isomeric opioids in the complex matrix of urine. Three groups of isomeric opiods and deuterated analogs were monitored at trace levels in human urine despite the possible matrix interferences from the urine. As with previous studies, retention time, CCS, and accurate mass were descriptive data used to identify and monitor the compounds of interest in the analyses. The high selectivity of the TIMS when paired with the LC and MS allowed for low detection levels comparable to liquid chromatography with tandem mass spectrometry (LC-MS/MS) and even with potentially less false-positives based on shared multiple reaction monitoring transitions between the isomers.

The opioids used in this study were as follows: 6-acetylmorphine (6-AM), naloxone, codeine, hydrocodone, morphine, hydromorphone, norcodeine, norhydrocodone, and the deuterated versions of each of these opioids (**Figure 5**). Calibration curves of the standard mixes and internal standard mixes were diluted with urine to create matrix match standards. This calibration curve in urine was created using liquid chromatography trapped ion mobility spectrometry with mass spectrometry (LC-TIMS-MS). LC separation was performed used a reversed-phase monolithic C18 column. Positive-mode electrospray ionization was used to ionize the opioids. The TIMS unit required specific parameters for gas flow and voltages at the entrance and exit funnels along with the mobility cell. An important parameter for the TIMS was the voltage at the exit funnel. As the voltage is decreased gradually at the exit funnel, the ions would exit based on the lowest mobility (largest CCS) first and the highest mobility (smallest CCS) last. Calibration standards for the instrument were used to calibrate voltage to CCS values. The theoretical CCS values of the opioids were calculated to predict the appropriate voltages needed to eject the ions from the mobility chamber.

The ion mobility of each compound was measured based on the calibration standards. 6-AM and naloxone had the highest CCS values of 176.7 and 171.1 $Å^2$. The orientation of the acetyl group on the 6-AM increased the CCS compared with the carbonyl group on the naloxone. The aliphatic group on the naloxone also extended out from the opioid body allowing the CCS to be larger compared with the rest of the opioids. In addition, computational modeling showed the tertiary amine having

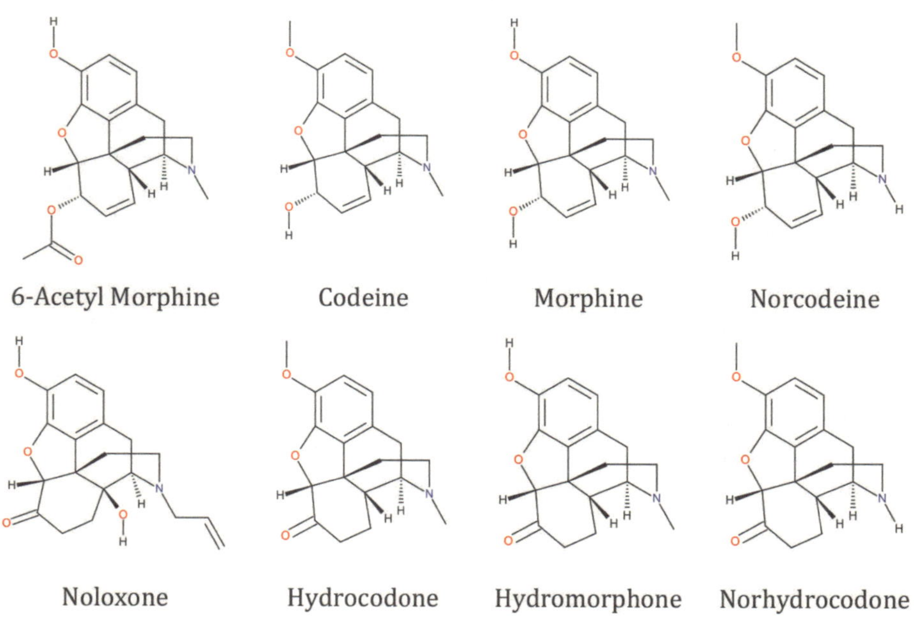

Figure 5.
Structures of opiod and opiod derivatives.

different orientations between 6-AM and naloxone. Even with this small difference in CCS values, the TIMS was able to separate these ions. Based on computational modeling, codeine and hydrocodone had CCS values of 168.2 and 167.8 Å^2. The hydroxyl versus carbonyl on carbon 6 was the only structural difference between these compounds, which led to small differences in CCS values and no separation in TIMS. There was no separation in the TIMS for morphine and hydromorphone due to the small difference in CCS values (162.9 and 163.3 Å^2) caused by small structural differences in the hydroxyl and carbonyl groups. Norcodeine and norhydrocodone were not separated in the TIMS due to the small difference in CCS values (167.9 and 167.4 Å^2) caused again by small structural differences in the hydroxyl and carbonyl groups. The opioid pairs of morphine/norcodeine and hydromorphone/norhydrocodone can be separated on TIMS due to a 5 Å^2 difference caused by a difference in secondary and tertiary amine orientation. The specificity in TIMS allowed for good peak resolution. The baseline mobility separations between 6-AM and naloxone, hydromorphone and norhydrocodone, and morphine and norcodeine were achieved with fast and slow scan TIMS. Despite the baseline separations of some opioid pairs, not all opioids were separated from each other in the calibration mix.

The study added liquid chromatography to the separation to obtain separation of the opioids with similar CCS values. Matrix interferences with water and urine in the TIMS with and without the LC were evaluated. Some matrix interferences could not be resolved with TIMS alone. Spiking the standards and internal standards into the urine and water showed increased limits of detection caused by interfering compounds. The liquid chromatography was successfully implemented to separate most matrix interferences with the standards and internal standards when run with TIMS. There were still slight LOD increases in urine using LC-TIMS-MS, but those increases were also seen while using LC–MS.

The LC runtime allowed for separation of the opioids. The internal standards of each opioid were deuterated and thus had a higher m/z when separated in the mass spectrometer despite having the same retention time and CCS values of their non-deuterated analytes. The internal standards were used as quality control checks of the retention times with the deuterated standards of each analyte containing different quantities of deuterium. Naloxone and 6-AM were successfully separated with LC and TIMS with retention times of 6.85 and 7.00 minutes. Hydrocodone and codeine were not successfully separated using TIMS, but the LC produced near-baseline separation with retention times of 6.8 and 7.0 minutes. Norhydrocodone and norcodeine had near-baseline resolution on the LC with retention times of 6.9 and 7.0 minutes despite having no separation in the TIMS. The retention times for morphine and hydromorphone were not mentioned in the LC-TIMS-MS analysis. Compounds with the same retention times can be separated in the TIMS based on previously mentioned CCS values. There was good reproducibility of CCS, LC retention times, and m/z values in between experiments and among different calibration levels of the calibration curve. The relative percent deviation (RPD) was <0.5% for CCS values with and without urine. Since the CCS values did not change based on concentration, this helped emphasize that CCS values can be used effectively for qualitative analysis in addition to retention time and accurate mass. Both additional data points showed low variability between water and urine, low and high concentrations, and inter-day performance. The LC-TIMS-MS protocol showed successful separation and quantitation for low ng/mL concentrations.

Combining LC with TIMS-MS improved peak resolution for instances when TIMS was not enough to separate the opioids based on very similar CCS values. In addition, the CCS values were very consistent with the drift tube ion mobility measurements and were consistent between experiments. This high reproducibility and specificity of the TIMS were very important when the opiods co-eluted in the LC dimension. Both the LC and TIMS techniques were crucial in order to separate compounds in this study.

5. Conclusions

This chapter has introduced the reader to basic ion mobility along with benefits and disadvantages of some ion mobility techniques. In addition, three studies were evaluated to show the benefits of coupling UPLC to ion mobility-MS techniques. This coupling allowed the separation of isobaric isomers that required an orthogonal separation. Descriptive data such as CCS and mass spectral fragmentation with drift time alignment helped guide structural confirmation and the formation of databases for tracking even if the compound identity was not confirmed. The drift time alignment also allowed for better quantitation of isobaric species. Ion mobility is not a perfect technique as seen in some of the above studies, but when paired with LC chromatography, the analytical strength of a method significantly increases.

Conflict of interest

The authors declare no conflict of interest.

Author details

Robert Owen Bussey III
RJ Reynolds Tobacco, Winston Salem, USA

*Address all correspondence to: busseyr@rjrt.com

References

[1] Ewing RG, Atkinson DA, Eiceman GA, Ewing GJ. A critical review of ion mobility spectrometry for the detection of explosives and explosive related compounds. Talanta. 2001;**54**:515-529. DOI: 10.1016/s0039-9140(00)00565-8

[2] Dodds JN, Baker ES. Ion mobility spectrometry: Fundamental concepts, instrumentation, applications, and the road ahead. Journal of the American Society for Mass Spectrometry. 2019;**30**:2185-2195. DOI: 10.1007/s13361-019-02288-2

[3] Mairinger T, Causon TJ, Hann S. The potential of ion mobility–mass spectrometry for non-targeted metabolomics. Current Opinion in Chemical Biology. 2018;**42**:9-15. DOI: 10.1016/j.cbpa.2017.10.015

[4] Ahrens A, Möhle J, Hitzemann M, Zimmermann S. Novel ion drift tube for high-performance ion mobility spectrometers based on a composite material. International Journal of Ion Mobility Spectrometry. 2020;**23**:75-81. DOI: 10.1007/s12127-020-00265-0

[5] Kirk A, Allers M, Cochems P, Langejuergen J, Zimmermann S. A compact high resolution ion mobility spectrometer for fast trace gas analysis. The Analyst. 2013;**138**:5200-5207. DOI: 10.1039/c3an00231d

[6] Shvartsburg AA, Smith RD. Fundamentals of traveling wave ion mobility spectrometry. Analytical Chemistry. 2008;**80**:9689-9699. DOI: 10.1021/ac8016295

[7] Morrison KA, Siems WF, Clowers BH. Augmenting ion trap mass spectrometers using a frequency modulated drift tube ion mobility spectrometer. Analytical Chemistry. 2016;**88**:3121-3129. DOI: 10.1021/acs.analchem.5b04223

[8] Tang K, Shvartsburg AA, Lee HN, Prior DC, Buschbach MA, Li F, et al. High-sensitivity ion mobility spectrometry/mass spectrometry using electrodynamic ion funnel interfaces. Analytical Chemistry. 2005;**77**:3330-3339. DOI: 10.1021/ac048315a

[9] Righetti L, Dreolin N, Celma A, McCullagh M, Barknowitz G, Sancho JV, et al. Travelling wave ion mobility-derived collision cross section for mycotoxins: Investigating interlaboratory and interplatform reproducibility. Journal of Agricultural and Food Chemistry. 2020;**68**:10937-10943. DOI: 10.1021/acs.jafc.0c04498

[10] Hamid AM, Ibrahim YM, Garimella SV, Webb IK, Deng L, Chen TC, et al. Characterization of traveling wave ion mobility separations in structures for lossless ion manipulations. Analytical Chemistry. 2015;**87**:11301-11308. DOI: 10.1021/acs.analchem.5b02481

[11] Morris CB, Poland JC, May JC, McLean JA. Fundamentals of ion mobility-mass spectrometry for the analysis of biomolecules. Methods in Molecular Biology. 2020;**2084**:1-31. DOI: 10.1007/978-1-0716-0030-6_1

[12] Giles K, Ujma J, Wildgoose J, Pringle S, Richardson K, Langridge D, et al. A cyclic ion mobility-mass spectrometry system. Analytical Chemistry. 2019;**91**:8564-8573. DOI: 10.1021/acs.analchem.9b01838

[13] Hernandez DR, Debord JD, Ridgeway ME, Kaplan DA, Park MA, Fernandez-Lima F. Ion dynamics in a trapped ion mobility spectrometer. The Analyst. 2014;**139**:1913-1921. DOI: 10.1039/c3an02174b

[14] Ridgeway ME, Lubeck M, Jordens J, Mann M, Park MA. Trapped ion mobility spectrometry: A short review. International Journal of Mass Spectrometry. 2018;**425**:22-35. DOI: 10.1016/j.ijms.2018.01.006

[15] Cumeras R, Figueras E, Davis CE, Baumbach JI, Gràcia I. Review on ion mobility spectrometry. Part 1: Current instrumentation. The Analyst. 2015;7:1376-1390. DOI: 10.1039/c4an01100g

[16] Kolakowski BM, Mester Z. Review of applications of high-field asymmetric waveform ion mobility spectrometry (FAIMS) and differential mobility spectrometry (DMS). The Analyst. 2007;**132**:842-864. DOI: 10.1039/b706039d

[17] Hatsis P, Kapron JT. A review on the application of high-field asymmetric waveform ion mobility spectrometry (FAIMS) in drug discovery. Rapid Communications in Mass Spectrometry. 2008;**22**:735-738. DOI: 10.1002/rcm.3416

[18] Yassin GH, Grun C, Koek JH, Assaf KI, Kuhnert N. Investigation of isomeric flavanol structures in black tea thearubigins using ultraperformance liquid chromatography coupled to hybrid quadrupole/ion mobility/time of flight mass spectrometry. Journal of Mass Spectrometry. 2014;**49**:1086-1095. DOI: 10.1002/jms.3406

[19] McCullagh M, Pereira CAM, Yariwake JH. Use of ion mobility mass spectrometry to enhance cumulative analytical specificity and separation to profile 6-C/8-C-glycosylflavone critical isomer pairs and known-unknowns in medicinal plants. Phytochemical Analysis. 2019;**30**:424-436. DOI: 10.1002/pca.2825

[20] Adams KJ, Ramirez CE, Smith NF, Muñoz-Muñoz AC, Andrade L, Fernandez-Lima F. Analysis of isomeric opioids in urine using LC-TIMS-TOF MS. Talanta. 2018;**183**:177-183. DOI: 10.1016/j.talanta.2018.02.077